EASEMENTS AND OTHER RIGHTS

Nick Isaac

and

Mark Walsh

Acknowledgment

Crown copyright material is reproduced with the permission of the Controller of HMSO and the Queen's Printer for Scotland.

Please note: References to the masculine include, where appropriate, the feminine.

Published by the Royal Institution of Chartered Surveyors (RICS)
Surveyor Court
Westwood Business Park
Coventry CV4 8JE
UK
www.ricsbooks.com

ISBN 978 1 84219 376 1

Typeset in Great Britain by Columns Design Ltd, Reading, Berks

Printed in Great Britain by Page Bros, Milecross Lane, Norwich

Printed on Greencoat Paper – Greencoat is produced using 80% recycled fibre and 20% virgin TCF pulp from sustainable forests.

Contents

Preface

This book, like the other titles in the Case in Point Series, is aimed at chartered surveyors and other professionals working in the construction and property industries.

While chartered surveyors may not need the breadth of knowledge of the law of their opposite numbers in the legal profession, they do need a similar depth of knowledge in those areas of the law which impinge directly on their work. Consequently, there are, for example, plenty of building surveyors who know more about the law relating to party walls, and plenty of quantity surveyors who know more about adjudication, than the average lawyer in general practice.

So surveyors need law, albeit in different specialist areas, according to the nature of their practice. This means that they need to maintain and develop their understanding of the law. The knowledge acquired at college, or in studying for the RICS Assessment of Professional Competence (APC), has a limited shelf life, and needs to be constantly updated to maintain its currency. Even the best practitioners (perhaps especially the best practitioners) are aware of the difficulty of keeping abreast of changes in the law. The most up-to-date specialist needs a source of reference as an aide-memoire or as a first port of call in more detailed research.

The Case in Point series

The books in the Case in Point Series are aimed at construction and other property professionals who need to upgrade or update their legal knowledge, or who need access to a good first reference at the outset of an inquiry (which is how lawyers also sometimes use them).

The series was established as part of the RICS Books commitment to meet the needs of surveying (and other) professionals. It was developed as a response to the particular difficulties created by the burgeoning of reported decisions of the courts. The sheer scale of the law reports, both general and specialist, makes it very hard to identify the significance of a particular decision, as part of a wider trend, as an isolated anomaly limited to its facts, or as a landmark departure in the law.

So it was decided to focus on developments in case law, although these are placed where necessary in the context of statutory or standard form contract provisions. In any given matter, the practitioner will want to be directed efficiently to the decision(s) bearing upon the issue with which he or she is dealing; in other words, to 'the case in point'. The authors have been selected as having the level of expertise required to be selective and succinct. The result is a high degree of relevance without sacrificing accessibility. The series has developed incrementally and now forms a collection of specialist handbooks which can deliver what practitioners need – the law on the matter they are handling, when they want it.

Easements and Other Rights by Nick Isaac and Mark Walsh

Knowing the identity of the registered owner of a building or a plot of land only tells part of the story, so far as the actual rights of enjoyment of the property are concerned. Of crucial importance in understanding the full picture are the rights and obligations which exist between owners and occupiers and this is true whether one is concerned with value, sale, development potential or simply normal use of the land/building. The principal focus of this latest addition to the Case in Point series is on easements, that is, rights enjoyed as part of the ownership of one property (the dominant tenement) over another property (the servient tenement). These are extremely wide-ranging and the cases in this book span rights of way, parking and unloading, drainage and other water rights, rights of support and protection of property and rights of light and air.

These matters are not only wide-ranging, they are often of considerable weight to the dominant and servient owners respectively. In a case like *Nickerson v Barraclough*, the extent of an implied easement of access could make the difference between being able, or not, to unlock the development potential of a site. Unsurprisingly, in an increasingly densely occupied country, many of the modern cases concern such matters as car parking, where space between neighbouring properties is at a premium. In the 2007 case of *Moncrieff v Jamieson*, the House of Lords had to consider whether a right to park was capable of being an easement ancillary to a right of way. Upon decisions like this depends not only the convenience of an individual owner, but in some cases the future saleability of the property.

Relationships between neighbouring properties and their owners are not solely governed by the common law of easements, and the authors have wisely chosen to extend the scope of their work to other related and comparable rights. An obvious example is the *Access to Neighbouring Land Act* 1992, by which owners can claim entitlement to enter upon their neighbours' property for the purposes of carrying out repair and maintenance to their own, where that is necessary. Another is the *Party Wall etc. Act* 1996, which regulates the relationship between owners of property sharing a common wall, fence or other shared structure (for which, see *Party Walls* by Sarah Hannaford and Jessica Stephens, also in the Case in Point Series). Yet another is the treatment of the technical area of what might be collectively termed 'utilities rights', namely telecommunications wayleaves and rights associated with water, gas and electricity.

As well as the substance of the rights themselves, the cases digested deal with their creation and extinguishment and with the tort of interference with easements and the remedies available, including injunctions, damages and self-help.

To review a body of cases which spans the four centuries between *Luttrel's case* in 1601 and *Wall v Collins* in 2007 requires both an impressive knowledge of the law and the capacity to sift out the landmark decisions from amongst a mass of fact-specific routine applications. RICS Books has been fortunate to secure the services for this title of two able London property lawyers.

Nick Isaac is a barrister at Tanfield Chambers, one of London's largest sets, specialising in property law. With 15 years' experience, Nick has appeared in a number of significant property cases, notably *Barton v Church Commissioners* in 2006.

Mark Walsh is also a barrister and formerly practised at Tanfield Chambers. He is now with the Guildford office of international law firm Clyde & Co LLP.

Anthony Lavers, 2008
Professional Support Lawyer, White & Case, London
Visiting Professor of Law, Oxford Brookes University
Consultant Editor, Case in Point Series

List of Acts, Statutory Instruments and abbreviations

The following Acts and Statutory Instruments are referred to in this publication. Where an Act or Statutory Instrument is mentioned frequently, it is referred to by the abbreviation in brackets that follows.

Access to Neighbouring Land Act 1992
Acquisition of Land Act 1981
Commons Act 2006
Commons Registration Act 1965
Communications Act 2003
Compulsory Purchase Act 1965
Countryside and Rights of Way Act 2000
Criminal Damage Act 1971
Electricity Act 1989
Gas Act 1986
Gas Act 1995
Highways Act 1980
Inclosure Acts
Land Registration Act 1925
Land Registration Act 2002 (**'LRA 2002'**)
Law of Property (Miscellaneous Provisions) Act 1989
Law of Property Act 1925 (**'LPA'**)
Local Government (Miscellaneous Provisions) Act 1976
London Building Acts
London Building Act 1894
London Building Acts (Amendment) Act 1939
National Parks and Access to the Countryside Act 1949

Party Wall etc. Act 1996
Perpetuities and Accumulations Act 1964
Prescription Act 1832
Rights of Light Act 1959
Telecommunications Act 1984
Utilities Act 2000
Water Industry Act 1991
Water Resources Act 1963
Water Resources Act 1991

Civil Procedure Rules Part 25
Civil Procedure Rules Practice Direction 56
Electricity (Compulsory Wayleaves) (Hearings Procedure) Rules 1967
Electricity Act 1989 (Consequential Modifications of Subordinate Legislation) Order 1990
Gas (Street Works) (Compensation of Small Businesses) Regulations 1996

The text of this publication is divided into commentary and case summaries. The commentary is enclosed between grey highlighted lines for ease of reference.

Table of cases

1 Introduction

1.1 EASEMENTS AND OTHER RIGHTS

This book deals with the creation, scope, use and extinguishment of easements and similar rights. It covers the legislation relevant to, as well as the decided cases which establish or illustrate key points in relation to such rights.

An 'easement' is a legally enforceable property right held by one person in his capacity as owner of one area of land ('the dominant tenement') over another area of land in the vicinity and in separate ownership ('the servient tenement') and which benefits the dominant tenement. Common examples of an easement are a right of way, a right of light or a right of support.

The extent to which the existence and extent of a particular easement affects the ability of a landowner to use his land is likely to be the primary concern of a surveyor dealing with easements. This book is therefore intended to give surveyors practical answers to the everyday questions: 'Is there an easement over this land?' and, 'How is this land affected by that easement?'. While it also answers some less straightforward questions, judgment may be required as to when a more comprehensive text, or even a specialist lawyer are called for.

Many of the cases referred to in the following pages arose at times of new industrial development, when watermills, canals and then railways came into being. While current developments are more likely to relate to high speed data cabling, the principles established in occasionally very old cases can still be applied to modern situations.

The other rights covered in this book are rights which are analogous to, but are nevertheless not easements: for example, the right to enter upon a neighbour's land under the *Access to Neighbouring Land Act* 1992, or the rights of riparian owners.

This book does not deal with profits à prendre, i.e. the right to take from the produce (game, fish, fruit etc.) or soil or minerals from the servient tenement, save to say that in general the rules for acquisition, extinguishment and disturbance of such profits are the same as for easements.

1.2 ESSENTIAL CHARACTERISTICS OF EASEMENTS

Re Ellenborough Park (1956)

A park was bordered on three sides by roads, and on a fourth by houses which enjoyed the benefit of 'the full enjoyment...in common with the other persons to whom such easements may be granted of [the park]'. In holding that this right amounted to an easement, the Court of Appeal confirmed that the four characteristics essential to an easement are:

- There must be a dominant and a servient tenement.
- An easement must accommodate the dominant tenement.
- Dominant and servient owners must be different persons.
- A right over land cannot amount to an easement unless it is capable of forming the subject matter of a grant.

1.2.1 The dominant tenement

Although the identity of the dominant tenement will usually be obvious, the court is permitted to look at the surrounding circumstances to determine which land is intended to form the dominant tenement.

Johnstone v Holdway (1963)

A conveyance of land adjoining a quarry reserved a right of way over that land for use 'at all times and for all purposes (including quarrying)', but did not identify the quarry as the dominant tenement. The Court of Appeal held that it was entitled to construe the conveyance by reference to 'evidence of all material facts at the time of execution of the deed' and had no hesitation in holding that the quarry was the dominant tenement.

1.2.2 The servient tenement

Woodman v Pwllbach Colliery Co Ltd (1914)

The Court of Appeal refused to imply an easement allowing a lessee to cause a nuisance by allowing coal dust to settle on neighbouring land, because the extent of the land over which the lessee claimed such a right was not a defined area. Swinfen Eady LJ said: 'It is necessary for an easement that there should be a servient tenement that can be defined and pointed out.'

1.2.3 Accommodating the dominant tenement

The servient tenement must provide a benefit to the land comprising the dominant tenement, rather than simply to the person owning the supposed dominant tenement.

Hill v Tupper (1863)

A lessee of land adjoining a canal was granted 'the sole and exclusive right or liberty to put or use boats...and to let the same...for the purposes of pleasure' on the canal. His claim against a defendant who also had premises on the canal failed on the basis that the exclusive right granted had no connection with the use and enjoyment of the lessee's land, and was unknown to the law.

Clapman v Edwards (1938)

A lessee of a petrol station was granted the right in his lease 'to use the flank walls of [adjoining premises] for advertising purposes'. The Court rejected an attempt to prevent the lessee permitting a third party to use the walls for advertising. The plaintiff argued that this was an easement which could not be assigned separately from the petrol station, but the Court held that the right of advertising was not connected with the petrol station and was therefore a licence not an easement.

1.2.4 Dominant and servient owners must be different persons

If a person owns two adjoining fields, Blackacre and Whiteacre, and walks across Blackacre to reach Whiteacre, he is merely exercising rights of ownership. Only if he sells or rents out Whiteacre is it possible for an easement over Blackacre to exist.

1.2.5 The right must be capable of forming the subject matter of a grant

A right to do almost any positive act on land is capable of forming the subject matter of a grant – a right to pass and re-pass (a right of way), or a right to park are obvious examples of such positive rights.

Negative rights, however, are far less likely to amount to a sufficient right. Rights of support or of light are the most common examples.

Hunter v Canary Wharf Limited (1997)

Residents of London's Docklands attempted to sue the owners of the 800-foot high Canary Wharf on the basis that its presence was interfering with their television signals. In holding that no nuisance had taken place, Lord Hoffman said that he did not consider that an easement of television could exist. He suggested that such a right, like a right to a view, would impose a burden on a very large and indefinite area; a burden too heavy for the law to support its imposition.

1.2.6 Rights amounting to joint or exclusive occupation are not easements

Copeland v Greenhalf (1952)

For a period of 50 years father and then son had used a strip of land separating a village and an orchard for the purpose of storing vehicles awaiting repair or collection, and for repairing vehicles. They always left an adequate access route to and from the orchard some 10 feet wide. Upjohn J found that this use amounted to a claim to possession of the servient tenement, either to the exclusion of the owner, or jointly with the owner. Consequently the right was too wide and undefined to constitute an easement (and should have been put as a claim to adverse possession).

1.3 OTHER RIGHTS

There are a number of other rights analogous or closely related to easements, which it is convenient to deal with in this book:

- Riparian rights, or rights relating to water, can be easements, but are more often simply part of the fee simple of land.
- Rights to access other land are provided by the *Access to Neighbouring Land Act* 1992.
- Fencing easements, so-called, stand on their own.

Egerton v Harding (1975)

This is the leading case on so-called 'fencing easements', where an occupier of land can be obliged to maintain a fence for the benefit of adjoining owners and occupiers. The claimant's cottage and defendant's farm both adjoined common land over which they both enjoyed grazing rights. The defendant's cattle strayed into the claimant's garden, but the claimant's claim for trespass was successfully defended on the basis of a finding that there was a custom obliging owners of land adjoining the common to fence against cattle lawfully on the common. The Court of Appeal approved this finding and also explained that such an obligation could originate in prescription or lost modern grant too.

2
Acquisition of easements

2.1 ACQUISITION GENERALLY

There are many ways in which an easement can be acquired. In this chapter we concentrate on the most common methods of acquisition.

Often, easements are acquired formally as part of a purchase or sale of land, and, where the land in question is registered, any easement over it should be recorded in the property register of the property's Land Registry entry. So, the Land Registry entries are the first place to check for the existence of easements, regardless of the method of acquisition. Conversely, easements over a servient tenement, at least if the subject of an express grant, will usually be shown in the charges register of the servient tenement's Land Registry entry.

Easements acquired informally are much less likely to be recorded in Land Registry entries, and it is to the land itself that you must look in assessing whether such easements do or might exist. Ask the owner/occupier about the use, if any, that adjoining owners/occupiers make of his land, and whether he makes any use of adjoining land ('Does anyone else use this access road/track/car park?'). Find out how long the use has continued, and whether there is anyone who can speak to that use from their own experience. If the use has continued for less than 20 years, it is unlikely that an easement will have arisen. If the use has gone on for longer, you will need to consider whether an easement has arisen by prescription.

2.2 EXPRESS GRANT

2.2.1 Grant by deed

The most common and straightforward express grant of an easement is a grant or reservation, contained in a deed – usually the conveyance effecting the transfer of land – by the freehold owner of one piece of land (the 'grantor'), in favour of the freehold owner of another piece of land ('the grantee').

The difference between a 'grant' and a 'reservation' is that in a grant the grantor gives a right to the grantee in addition to the land he is transferring, whereas in a reservation the grantor reserves or retains a right for his own or others' use over the land he is transferring.

Since the *Law of Property (Miscellaneous Provisions) Act* 1989 came into force the formalities required to create a deed are minimal. A deed must:

- describe itself as a deed – the heading 'Deed' or the appellation 'signed as a deed' both suffice;
- be signed by the parties to it – in the case of a deed creating an easement, the parties would be grantor and grantee;
- have those signatures witnessed. For a signature to be properly witnessed, the attesting witness should actually watch the original signature being written, then add 'witnessed by', their own signature, name and address. There is no reason why grantor and grantee cannot witness each others' signatures;
- be 'delivered'. This does not necessarily mean the physical transfer of the document (although it may do), but merely that the transaction is irreversible. Words on the face of the deed to the effect that it is deemed delivered will suffice.

If and to the extent that a deed is effecting a transfer of land it will generally be required to be registered at HM Land Registry. A deed transferring land will require registration for the transfer to be complete. Equally a deed simply creating an easement, in order to 'operate at law' requires to be completed by registration at the Land Registry. The effect of failing to register such a transfer or easement is that it is only effective in equity, and so will not bind a bona fide purchaser for value without notice of it.

2.2.2 Grant by will

There are two ways in which a testator can create an easement by will:

1. where he owns and makes separate dispositions of the dominant and servient tenements;
2. where he owns land and grants a new easement in favour of land not owned by him.

2.2.3 Section 62 of the Law of Property Act 1925

Section 62 of the *Law of Property Act* 1925 provides as follows:

> '(1) A conveyance of land shall be deemed to include and shall by virtue of this Act operate to convey, with the land, all buildings, erections, fixtures, commons, hedges, ditches, fences, ways, waters, watercourses, liberties, privileges, easements, rights, and advantages whatsoever, appertaining or reputed to appertain to the land, or any part thereof, or, at the time of the conveyance, demised, occupied, or enjoyed with, or reputed or known as part or parcel of or appurtenant to the land or any part thereof.
>
> (2) A conveyance of land, having houses or other buildings thereon, shall be deemed to include and shall by virtue of this Act operate to convey, with the land, houses, or other buildings, all outhouses, erections, fixtures, cellars, areas, courts, courtyards, cisterns, sewers, gutters, drains, ways, passages, lights, watercourses, liberties, privileges, easements, rights and advantages whatsoever, appertaining or reputed to appertain to the land, houses, or other buildings conveyed, or any of them, or any part thereof, or, at the time of the conveyance, demised, occupied, or enjoyed with, or reputed or known as part or parcel or appurtenant to, the land, houses, or other buildings conveyed, or any of them, or any part thereof.
>
> (3) A conveyance of a manor shall be deemed to include and shall by virtue of this Act operate to convey, with the manor, all pastures, feedings, wastes, warrens, commons, mines, minerals, quarries, furzes, trees, woods, underwoods,

coppices, and the ground and soil thereof, fishings, fisheries, fowlings, courts leet, courts baron, and other courts, view of frankpledge and all that to view of frankpledge doth belong, mills, mulctures, customs, tolls, duties, reliefs, heriots, fines, sums of money, amerciaments, waifs, estrays, chief-rents, quit-rents, rentscharge, rents seek, rents of assize, fee farm rents, services, royalties, jurisdictions, franchises, liberties, privileges, easements, profits, advantages, rights, emoluments, and hereditaments whatsoever, to the manor appertaining or reputed to appertain, or, at the time of the conveyance, demised, occupied, or enjoyed with the same, or reputed or known as part, parcel, or member thereof...

(4) This section applies only if and so far as a contrary intention is not expressed in the conveyance, and has effect subject to the terms of the conveyance and the to the provisions therein contained.

(5) This section shall not be construed as giving to any person a better title to any property, right, or thing in this section mentioned than the title which the conveyance gives to him to the land or manor expressed to be conveyed, or as conveying to him any property, right or thing in this section mentioned, further or otherwise than as the same could have been conveyed to him by the conveying parties.'

Conveyances used to include a lengthy list of general rights similar or identical to the list now contained in section 62 of the *Law of Property Act* 1925. That section, and its predecessor, removed the need for such a list. Do not be put off by the length of the list and the terms used in it. There are very, very few lawyers who would be able to tell you what a 'frankpledge' or an 'amerciament' is. The reason for including section 62 in this volume is that is does occasionally create easements, often unintentionally, on the conveyance of land. Easements are created by section 62 where, at the time of the conveyance, rights etc. are enjoyed, by an occupier of the land conveyed, over other land of the grantor.

Wright v Macadam (1949)

An existing tenant of a flat used a shed in the adjoining garden for storing coal, with the permission of his landlord. That permission could have been withdrawn at any time.

However, when the landlord granted a new lease to several tenants, one of whom was the existing tenant, the Court of Appeal held that section 62 operated to grant, as part of the new tenancy, a right to use the shed for the purpose of storing coal, together with the necessary means of access to it. This case also illustrates that 'conveyance' for the purposes of section 62 includes a lease.

Regis Property Co. Ltd v Redman (1956)

The tenant of a flat claimed a right to the supply of hot water and central heating from his landlord under section 62. The Court of Appeal head that an obligation to supply constant hot water or central heating was essentially a matter of personal contract, and was not a right, easement or privilege capable of being granted by lease or conveyance so as to pass under that section.

Ward v Kirkland (1967)

A farmyard and cottage had been in common ownership. During that period the occupier of the cottage had always asked for and received permission to access the farmyard for the purpose of maintaining the wall of the cottage. Once ownership was split, the plaintiffs were able to establish that they had a right of access for that purpose by virtue of section 62.

P & S Platt Ltd v Crouch (2003)

The defendants were the owners of a hotel, a house in the hotel grounds and a nearby area of land bordering a river on which was built a bungalow. The claimant and defendant entered into an agreement for the claimant to purchase the hotel business from the defendant, with the claimant and defendant each having the rights over the other's land that they would have had were they two separate buyers to whom the defendant had made simultaneous transfers of the properties and the retained land. The transfer did not indicate any intention to exclude or modify the application of section 62. The Court of Appeal held that the evidence was clear that the mooring and fishing rights in question were part of the hotel business and advertised as such and enjoyed by hotel guests, as well as being clear and apparent. Section 62 therefore operated to convert the rights into full easements. Although the easements did detract from the defendants' enjoyment of their land, it was not a sufficiently substantial interference to prevent the rights from being easements.

2.3 IMPLIED GRANT

Easements can also be created by implication, usually on the transfer of land, but also on the grant of a lease or under a gift by will. The question is always whether an intention to grant the easement in question can properly be inferred, and there are several different circumstances in which this can occur:

- under the doctrine of non-derogation from grant;
- under the rule in *Wheeldon v Burrows*;
- from the description in the parcels clause;
- from the particular use intended for the land;
- of necessity.

2.3.1 The doctrine of non-derogation from grant

The doctrine of non-derogation from grant, as it applies to the creation of easements, is generally limited to the relationship of landlord and tenant. It operates to prevent a landlord who has retained land from doing something on that land which would make the premises let to his tenant unfit or materially less fit to be used for the purpose for which the letting was made.

Hilton v James Smith and Sons (Norwood) Ltd (1979)

A landlord had three commercial units with a service road to the rear. The units were let to tenants, each of whom covenanted not to allow their vehicles to obstruct the service road. The tenant of the end unit successfully sued the landlord in nuisance and for derogation from grant because the landlord failed to enforce that covenant against the other lessees, whose breaches prevented the tenant from using the service road for his delivery vehicles.

Johnston & Sons Ltd v Holland (1988)

The defendant was long lessee of an advertising hoarding site which could be seen from the road over an intervening piece of open land. At the date of grant the plaintiff's predecessor in title did not own the open land, but acquired it later, and put up a hoarding on that land which obscured the advertising hoarding. The Court of Appeal held that this amounted to a derogation from grant on the part of the plaintiff. Nicholls LJ usefully summarised the broad principle of non-derogation from grant as: '...if one man agrees to confer a particular benefit on another, he must not do anything which substantially deprives the other of the enjoyment of that benefit; because that would be to take away with one hand what is given with the other'.

Platt v London Underground Ltd (2001)

The claimant took a lease of a kiosk at one of the exits of Goodge Street underground station on the understanding that the exit would be open when the station was. In fact the exit was only open during the morning rush hour. The Court held that the closure of the exit at other times constituted a derogation from grant.

2.3.2 The rule in Wheeldon v Burrows

Although generally cited as a separate common law rule, the rule in *Wheeldon v Burrows* is really only another example of non-derogation from grant. It grants easements by implication when the owner of a piece of land grants away part of it. It applies to those easements necessary to the reasonable enjoyment of the property granted which have been used and are, at the time of the grant, used by the entirety of the property for the benefit of the part granted. Obvious examples are paths and drains.

Wheeldon v Burrows (1879)

A piece of land and adjoining workshop were sold by the same vendor one month apart. The windows of the workshop looked out onto and received their light from the piece of land. The Court held that there was no implied reservation of a right to light for the workshop. NB this was actually a case about implied reservation, but the Court's judgment stated the rules applying to implied grant as well.

Borman v Griffith (1930)

A drive ran past what had formerly been a gardener's lodge, to the hall. The owner let the lodge to the plaintiff, then let the hall to the defendant, who promptly blocked the drive. In giving judgment for the plaintiff, Maugham J said that 'where, as in the present case, two properties belonging to a single owner and about to be granted are separated by a common road, or where a plainly visible road exists over the one for the apparent use of the other, and that road is necessary for the reasonable enjoyment of the property, a right to use the road will pass with the quasi-dominant tenement, unless by the terms of the contract that right is excluded'.

Millman v Ellis (1996)

A conveyance of land granted to the purchaser a right of way over part of a layby which gave access to a highway. However, although access by the express right of way was possible, it was dangerous. In these circumstances the Court held that the purchaser was also entitled to a right of way over the remainder of the layby. The use of all of the layby for the purpose of gaining access to the highway was continuous and apparent, and was reasonably necessary in order to obtain safe access to the highway.

2.3.3 Description in the parcels clause

In a few cases the Court has found it possible to infer an intention to grant an easement from the language of the conveyance or lease.

Espley v Wilkes (1872)

The defendant's lease described the demised premises as 'bounded on the east and north by newly-made streets', which streets were shown on the lease plan. The street to the east was never in fact made or marked out, and was susbsequently leased by the same landlord to the plaintiff. The Court held that the effect of the defendant's lease was to give him a private right of way over both streets, the landlord being estopped by his own description from denying that there were streets there.

Rudd v Bowles (1912)

Four new houses had gardens to the rear, which adjoined the lessor's land. Each garden had a gate leading on to the lessor's land, and, apart from these gates, the only access to the gardens was through the houses. The lease plans for the houses showed a coloured strip on the lessor's land, but it was not mentioned in the body of the leases. In these circumstances the Court held that each lease contained the implied grant of a right of way over the strip.

2.3.4 From the particular use intended

Lyttleton Times Co. Ltd v Warners Ltd (1907)

A lease had been negotiated on the basis that a building would be rebuilt, the lessee using the upper floors of the rebuilt building for hotel bedrooms, the lessor using the lower floors as a printing works. An action by the lessee to restrain a noise nuisance by the landlord (in using his printing works in a normal way) failed because the Court held that the use of the lower floors as a printing works was exactly what had been contemplated by the parties, and the landlord must therefore be taken to have impliedly reserved the right to use the premises in that way, even though it caused a nuisance.

Pwllbach Colliery Co. v Woodman (1915)

A lease which recognised that the lessee would or might carry on mining operations did not impliedly grant the right to create a nuisance by dissipating coal dust over the lessor's land by the use of screening plant installed after the grant, since there was no evidence that mining could not be carried out at the site without use of the screening plant.

Stafford v Lee (1992)

A transfer of woodland did not contain an express right of way over a private roadway which provided the only practical access. The parties agreed that a right of way was implied, but disagreed as to whether it extended to the use necessary for building and using a house on the land. The Court of Appeal inferred from the plan attached to the transfer that the parties intended that the plot should be used for the construction of a house, and that a right of way for the building and use of such house should be implied. Nourse LJ set out the two hurdles a person claiming such an implied easement must establish, namely: (1) a common intention that the land is to be put to some definite and particular use, and (2) that the easements claimed are necessary to give effect to that use.

Davies v Bramwell (2007)

Land had been transferred for use as a commercial garage with a hydraulic ramp. In order for larger vehicles to have reasonable access to the ramp, it was necessary for them to cross a section of the retained land, over which no express right of way was granted. The Court of Appeal held that this intended use (which was in fact carried out for several years after grant before a dispute arose) was sufficient to give rise to create an implied right of way.

2.3.5 Of necessity

An easement implied as a matter of necessity is very rare indeed, and applies only to rights of way. It arises where an owner grants away part of his land, and either the part granted away or the part retained is left without any legally enforceable means of access. In those circumstances the otherwise inaccessible part is entitled, as of necessity, to a right of way on foot over the other part. It should be emphasised that *any* other access will prevent the creation of a way of necessity.

Often in practice it is more fruitful to consider whether a right of way has been created by implication as a consequence of the doctrine of non-derogation from grant, or in the light of the parties' intention as to the use of the dominant land, since this is more likely to be a source of the required right.

Bolton v Bolton (1879)

A right of way of necessity arose in favour of a school board. Two potential ways existed. The Court held that where a right of way of necessity arises, it is for the grantor to choose the line of the way.

Nickerson v Barroughclough (1980)

This is the leading modern case on easements of necessity. In this case the Court of Appeal held that the doctrine of way of necessity is founded upon an implication from the circumstances of a grant of land, and was not a free-standing rule of public policy

Manjang v Drammeh (1990)

The land claiming a right of way by necessity enjoyed no access whatsoever over adjacent land. It did, however, have an available access by water. In those circumstances, the Privy Council held that the access by water was sufficient to prevent any implication of a way of necessity.

2.4 ACQUISITION BY PRESCRIPTION

It is a fundamental principle of property law that land should not be allowed to stagnate. The law of adverse possession and the law of prescription are both examples of where the law steps in to formalise *de facto* ownership or use of land where the same has continued over a lengthy period of time. This section deals with the circumstances in which a particular use of property over a period of time can give rise to easements. As a handy rule of thumb, if a quasi-easement has been exercised for 20 years or longer without permission, an easement is likely to have been created.

There are a number of requirements common to almost any claim to prescription, which can be summarised as follows:

- The enjoyment of the right must generally be against the freehold owner.
- The enjoyment of the right can be by a freehold or leasehold owner.
- The enjoyment must be as of right, i.e. enjoyed openly, without permission, and without force.
- The servient owner must know about the enjoyment of the right, or have the means of knowledge.
- The servient owner must have been able to (but did not) interrupt enjoyment of the right.
- Enjoyment must be definite and continuous in nature.

- Enjoyment must be consistent with common law and statute.

There are also three different methods of acquiring rights by prescription, which, going from the least to the most common, are:

- common law prescription;
- the doctrine of lost modern grant;
- under the *Prescription Act* 1832.

2.4.1 Enjoyment against the freehold owner

Pugh v Savage (1970)

The defendant tenant counterclaimed a right of way over a lane and field giving access to fields he had rented from the plaintiff, on the basis that he and his predecessors had used that right for the previous 30 years. It was agreed between the parties that the servient land had been let out for a period of some ten years during the period of use. The Court of Appeal held that where a tenancy came into existence during the period of user relied upon, the grant of the tenancy would not, in the absence of evidence that the servient owner had no knowledge of such user, be fatal to the presumption of a grant or to a claim made under the *Prescription Act*, although it was a factor to be considered. This was to be distinguished from a case where the use commenced after the grant of a tenancy, because in those circumstances the servient owner might not have been able to stop the user even if he knew about it.

2.4.2 Who has to enjoy the right?

A freehold owner who enjoys a right obviously has sufficient title to make a claim to a prescriptive right. However, a tenant's possession of land is treated by the law as the landlord's possession also, so it too will suffice.

2.4.3 Enjoyment as of right

This is generally the most problematic area of the law relating to the acquisition of easements by prescription, and calls for several examples and explanations.

ceased to be permissive from the time when the plaintiff started using the way regularly, and stopped asking for permission.

Ratique v Walton Estates (1993)

The owner of a way put up a sign saying that anyone who used the way did so by permission. This was held to be sufficient to stop time running in a claim for a prescriptive right to use the way.

2.4.4 Extent of use required

Although use has to continue for a long period of time in order to give rise to a claim for prescription, the use does not need to be very extensive to give rise to a right. It should always be remembered, however, that the extent of right obtained by prescription will be determined by the extent of the use made in acquiring prescription.

Monmouthshire Canal Co. v Harford (1834)

Parke B. said: 'An enjoyment of an easement for one week and a cessation to enjoy it during the next week and so on alternately, would confer no right.' It should be noted that this would not be the case if the claimant did not use the quasi-easement simply because he had no call to use it.

Diment v Foot (1974)

The use of a way six to ten times a year for agricultural purposes was held to be sufficient use to give rise to a right of way by prescription.

2.4.5 Servient owner must know about enjoyment

The servient owner must know about the enjoyment, or have the means of finding out about the enjoyment, if a right by prescription is to be established. The most common difficulty in this respect arises where the servient owner has let out his land for the period of use relied upon by the dominant owner. Generally, if the land has been let during that period, the servient owner will not be considered to have sufficient knowledge or means of knowledge to give rise to a prescriptive right.

Davies v Du Paver (1953)

A right of sheepwalk (i.e. to walk sheep over the servient owner's land) was claimed by prescription. The land over which the right was claimed had been let out during all of the 60 years of use, except for a few years at the end of that period. The Court held that it was for the plaintiff to show that the owner had some knowledge, or reasonable means of knowledge, of the user. Although the plaintiff was able to establish common local knowledge of the user, this was held to be insufficient to extend to the owner, who lived several miles away and did not necessarily share that knowledge.

2.4.6 Servient owner must be able to interrupt the user

Sturges v Bridgman (1879)

Two houses stood next to one another, the first belonging to confectioners, the second to a doctor. For more than 60 years the confectioners had used a pestle and mortar in their kitchen, causing noise. The doctor had only bought the neighbouring house a few years before action, and built a consulting room close to the confectioners' kitchen. In an action brought by the doctor, the confectioners contended that they had acquired a right to cause the noise (which was found to constitute a nuisance) by prescription. The Court held that no such right had been obtained because, prior to the building of the consulting room, the owner of the doctor's house could not have prevented the use by legal action, and therefore there was insufficient use from which a grant could be presumed.

2.4.7 Enjoyment must be definite and continuous in nature

Hulley v Silversprings Bleaching and Dyeing Co. Ltd (1922)

A mill owner claimed a prescriptive right to pollute a stream. However, during the period relied upon to gain the prescriptive right, the mill had steadily increased the volume of water polluted (as it gradually increased its plant). The

Court held that this increase was 'destructive of that certainty and uniformity essential for the measurement and determination of the user by which the extent of the prescriptive right is to be ascertained'.

White v Taylor (No. 2) (1969)

The plaintiff was claiming various sheep rights (pasturage, watering and bringing down) under section 1 of the *Prescription Act* 1832. The Court held that the plaintiff did not need to establish that the right had been exercised continuously, since the nature of the right was that it would only be used intermittently. However, the use must still be shown to have been of such a character, degree and frequency as to indicate an assertion by the plaintiff of a continuous right and of a right of the measure of the right claimed. The Court held in this case that the evidence did not establish a sufficient length of use.

2.4.8 Right must be consistent with statute and custom

Prescription is based upon a presumption that a grant has been made. It is impossible to make such a presumption where such a grant would have been contrary to a statute or custom. However, a grant can still be presumed where one could have been made, but where the use was unlawful in the meantime.

Cargill v Gotts (1981)

The plaintiff claimed, by prescription, a right to take water for irrigation of a farm from a mill pond forming part of a river. For part of the period relied upon the user was illegal under section 23(1) of the *Water Resources Act* 1963. Consequently the Court held that any period during which the use was illegal could not be relied upon by the plaintiff, since 'the court will not recognise an easement established by illegal activity'.

Bakewell Management Ltd v Brandwood (2004)

The defendant appealed against a decision that he was not entitled by reason of more than 20 years' uninterrupted use to a right of way for vehicles across a common owned by the claimant. The defendant, who needed to use the common to gain access to the nearest public road, accepted that neither the claimant nor its predecessors in title had given permission for the common to be used for vehicular access. Although it is an offence under the *Law of Property Act* 1925, section 193(4) to drive without lawful authority on a common, and despite no lawful authority having ever been given to drive on the common, the House of Lords held that the defendant did acquire a right (by lost modern grant), since a right of way could have been granted by the claimant or its predecessor in title which would have made the use lawful. Only where the grantor could not lawfully make the grant would it be impossible for a prescriptive right to arise.

2.4.9 Prescription at common law

At common law one has to show that a right has been exercised since 1189 (the accession of Richard I) in order to establish a right by prescription. Unsurprisingly, this is almost impossible to do in practice. Although evidence of use for the period of living memory raises a presumption of use since 1189, evidence that the use in fact commenced later than 1189 suffices to rebut that presumption.

Lost modern grant

Given the obvious difficulties presented by common law prescription, the courts developed the legal fiction of the lost modern grant. The fiction presumes that if someone has been exercising a right for over 20 years, then the owner of the quasi-servient tenement must have granted a right for such use, but that the deed of grant has in the meantime been lost.

Most presumptions are rebuttable, which is to say that they can be defeated by evidence to the contrary. Not so lost modern grant. Evidence that a grant was not in fact made will not rebut the presumption. In fact the only evidence capable of rebutting the presumption is evidence as to the impossibility of a grant – e.g. where the grantor lacked capacity to make such a grant.

Despite the existence of the *Prescription Act* 1832 (discussed below at paragraph 8.2), the doctrine of lost modern grant does have a place in modern law. As long as the use in question can be shown to have continued for *a* period of 20 years, it does not matter that the use ceased prior to the commencement of proceedings. Use down to the date of proceedings is, however, a requirement of the *Prescription Act*.

Dalton v Angus (1881)

The plaintiffs proved enjoyment of support for their factory for a period of 27 years before the accident which gave rise to the action. They claimed a right of support under the doctrine of lost modern grant, but it was proved or they admitted at trial that no grant had ever in fact been made. After an inordinate amount of judicial input (seven High Court judges were asked for their opinion on the point to assist the House of Lords), the effect of the House of Lords' decision was that the presumption of a lost modern grant was not rebuttable by evidence that no grant had ever been made, but could only be rebutted by evidence of incapacity on the part of the presumed grantor.

2.4.10 Prescription Act 1832

Section 2 of the *Prescription Act* 1832 provides as follows:

> 'No claim which may be lawfully made at the common law, by custom, prescription or grant, to any way or other easement, or to any watercourse or the use of any water to be enjoyed or derived upon, over, or from any land or water of our said lord the King, or being parcel of the Duchy of Lancaster or of the Duchy of Cornwall, or being the property of any ecclesiastical or lay person, or body corporate, when such way or other matter as herein last before mentioned shall have been actually enjoyed by any person claiming right thereto without interruption for the full period of twenty years, shall be defeated or destroyed by showing only that such way or other matter was first enjoyed at any time prior to such period of twenty years, but nevertheless such claim may be defeated in any other way by which the same is now liable to be defeated; and where such way or other matter as herein last before mentioned shall have been so enjoyed as aforesaid for the full period of forty years, the right thereto shall be deemed absolute and indefeasible, unless it shall appear that the same was enjoyed by some consent or agreement expressly given or made for that purpose by deed or writing.'

Section 2 almost certainly applies to all rights capable of being easements, save for rights of light, which are dealt with separately in section 3 of the Act, and which are discussed below.

The *Prescription Act* introduced new periods in relation to the acquisition of easements, but does not change the common law requirements as to the quality of use required to give rise to a prescriptive right. The use must still be as of right and enjoyed as an easement.

A claim based on 20 years' use is theoretically more easily defeated than a claim based on 40 years' use, since a 20-year claim can fail for various reasons, while a 40-year claim can only be defeated by evidence of a written consent or agreement. In practice, a claim is very rarely brought on the basis of the 40-year rule, since parties can generally rely more easily on the doctrine of lost modern grant.

The most common obstacle to the successful acquisition of rights under the *Prescription Act* is the fact that, by reason of section 4 of the Act, the right must have been enjoyed 'next before some suit or action'. So, to acquire a right under the Act, the dominant owner must have enjoyed the right for the 20 or 40 years immediately before an action is commenced.

While an interruption of less than a year does not count for the purposes of this requirement, evidence of several interruptions of less than a year may be considered by the court to demonstrate contentious user (i.e. user not 'as of right') and prevent acquisition entirely, or may result in the acquisition of a more limited right.

Finally in relation to the periods under the Act, it should be noted that section 7 of the *Prescription Act* provides for the exclusion of certain periods of time in the computation of the 20- or 40-year period required to acquire a right. Any periods during which 'any person capable of resisting any claim...shall have been or shall be an infant, idiot, *non compos mentis, feme covert,* or tenant for life, or during which any action or suit shall have been pending, and which shall have been diligently prosecuted, until abated by the death of any party or parties thereto' are excluded.

Section 3 of the *Prescription Act* 1832 provides as follows:

> 'When the access and use of light to and for any dwelling-house, workshop or other building shall have been actually enjoyed therewith for the full period of twenty years without interruption, the right thereto shall be deemed absolute and indefeasible, any local usage or custom to the contrary notwithstanding, unless it shall appear that the same was enjoyed by some consent or agreement expressly made or given for that purpose by deed or writing.'

Most of the points mentioned in relation to section 2 also apply to section 3 of the *Prescription Act*. However, there are some additional points and differences to be noted.

There is no 'as of right' requirement in relation to section 3, since one simply has to erect a building with windows or other openings in order to enjoy a putative right of light.

Section 3 does not bind the Crown.

It should be noted that the consent or agreement in writing is often contentious, and any such document should be analysed carefully.

Mitchell v Cantrill (1887)

A lease demised land 'except rights, if any, restricting the free use of any adjoining land or the conversion or appropriation at any time thereafter of such land for building or other purposes, constructive or otherwise'. The Court held that this amounted to a reservation from grant, but did not constitute consent or agreement within section 3.

Haynes v King (1893)

A lease contained a declaration 'notwithstanding anything herein contained, the lessors shall have power, without obtaining any consent from, or making any compensation to, the lessee, to deal as they may think fit with any of the premises adjoining or contiguous to the hereditaments hereby demised and to erect, or suffer to be erected, on such adjoining or contiguous premises, any buildings whatsoever, whether such buildings shall or shall not affect or diminish the light or air which may now, or at any time during the term hereby granted, be enjoyed by the lessee, or the tenants or occupiers of the hereditaments hereby demised'. The

Court held that this amounted to a consent or agreement within section 3 which prevented a right of light being acquired by prescription.

2.5 ACQUISITION BY ESTOPPEL

In addition to the methods of acquisition mentioned above, it is also possible to acquire an easement by way of proprietary estoppel. Strictly speaking such a right is not a legal right, but an equitable right. A discussion of the history and distinction between such rights is beyond the scope of this book. However, the practical application of the principles of proprietary estoppel in so far as they affect easements can be dealt with here.

The essential ingredients of proprietary estoppel as it relates to easements can be summarised as follows:

- an owner of land induces or encourages or allows another person to believe that he has or will have some right over the owner's land;
- that other person acts to his detriment in reliance upon that belief;
- the first landowner knows about that act of detrimental reliance; and
- the landowner then seeks to rely on his strict legal rights (which do not include the rights the other person believes himself to have or be entitled to).

In those circumstances the Court will make a declaration or order which will have the effect of forcing the landowner to give effect to the right which the other person believed he had or would obtain. This is known as 'giving effect to the equity'. In some cases this will simply mean that the Court will refuse the landowner an injunction preventing the other person, for example, using a right of way. In other cases, as appropriate, the Court may order the landowner to enter into a deed of grant in relation to the right in question. In both cases, the practical effect of the proprietary estoppel is to create an easement in favour of the other person.

In practice, proprietary estoppel often comes to the rescue of informal agreements which have not been reduced into writing, but have been acted upon by the parties thereto. An obvious example in practice would be where A agrees orally to a right of way over his land in favour of B. B then spends money improving the way and/or creating a driveway which leads to it. In those circumstances an estoppel would undoubtedly arise.

Ward v Kirkland (1967)

The owner of a farm was asked for and gave permission for the plaintiff to install drains through the farmyard. No stipulation was made by the farm owner as to the period of permission. The plaintiff laid the drains accordingly. When, some three years later, the new owner of the farm purported to terminate the permission, the Court held that she was not entitled to do so because of the plaintiff's expenditure in reliance upon the open-ended permission.

E.R. Ives Investment Ltd v High (1967)

The defendant was a builder who built a house on a site in 1949. At the same time the adjoining owner was developing flats on its land. The foundations of those flats encroached onto the defendant's land. At a meeting between the parties, the defendant and adjoining owner agreed and later recorded in correspondence that the foundations could stay but that he should have a right of way across the yard of the flats. Several years later, having used the way in the meantime, the defendant built a garage to which the only means of access was across the yard. The then owners of the flats did not object, and indeed persuaded the defendant to pay one-fifth of the cost of resurfacing the yard. In 1962 the plaintiff bought the flats and sought an injunction preventing the defendant from trespassing on the yard. The Court of Appeal held that the defendant had in equity a good right of way across the yard. One of the bases for that finding was proprietary estoppel.

Crabb v Arun District Council (1976)

The plaintiff owned a piece of land from which access to the highway was at point A over a road belonging to the defendant. The plaintiff wanted to sell his land in two parts and agreed with the defendant's representative that he could have a second access at point B. The plaintiff sold the piece of land with access at point A without reserving any right in favour of the remaining piece of land. The defendant then removed a gate it had installed at point B and fenced the gap. The Court of Appeal held that the plaintiff had acted to his detriment (in selling the first piece of land without reserving any right to use point A) in reliance upon the agreement with the defendant, and in those circumstances, to give effect to the equity of the situation, the defendant was entitled to an easement providing access at point B.

2.6 ACQUISITION BY STATUTE

Historically it has been reasonably common for easements to be created by statute – for example in relation to the *Inclosure Acts*, and various Acts relating to railways.

While it is always possible for Parliament to enact new legislation which might have the express or implied effect of creating easements, there is little current legislation which has that effect.

Section 13 of the *Local Government (Miscellaneous Provisions) Act* 1976, as amended, provides for the acquisition, by compulsory purchase order, of rights which are not in existence when the order is made.

3 Extinguishment

3.1 EXPRESS RELEASE

An easement can be extinguished by express release. Such release must be by deed to be valid in law. The requirements for a valid deed were set out in Chapter 2 at paragraph 2.2.1.

However, an agreement in writing and complying with the other requirements of section 2 of the *Law of Property (Miscellaneous Provisions) Act* 1989 can be specifically enforced, by the Court ordering the owner of the dominant tenement to enter into a deed of release.

Further, acquiescence and estoppel apply to the release of an easement in the same way they apply to the grant of an easement.

3.2 IMPLIED RELEASE

If an easement is granted other than for a specific term (usually as part of a lease), the right is perpetual. Consequently, only where a person entitled to the easement has demonstrated a fixed intention never at any time thereafter to assert the right himself, or to transfer that right to someone else, will the Court be prepared to find that there has been implied release of the easement.

Whether or not the facts of a particular case may give rise to an implied release, one must first decide whether the easement in question is a continuous easement (such as a right of light), or a discontinuous one (such as a right of way).

3.2.1 Implied release of continuous easements

Although there is no specific period of cessation which the authorities fix as sufficient to demonstrate an intention to release the easement, there is no reason in principle why it should be a period of many years.

If the dominant owner changes the use of his land so as to impose an additional burden on the servient land, this *may* destroy the continuous easement altogether.

If acting for the dominant owner of a right of light who intends to redevelop an existing building, it is essential that an accurate record is obtained of the exact size and location of the windows in the original building which enjoy that right of light. By doing so, the dominant owner can ensure that the new windows are positioned so as to preserve the existing rights of light, and rebut any assertion that the right has been abandoned.

Luttrel's Case (1601)

The plaintiff owned two fulling mills which enjoyed a right to a flow of water from a watercourse. The plaintiff pulled down the original decrepit mills, and built two grist mills in their place. The Court held that the alteration from fulling to grist mills was one of quality not substance, and that it was not prejudicial to the owner of the watercourse, so that the right to the flow remained.

Saunders v Newman (1818)

The owner of a watermill altered the dimensions of the mill-wheel. In deciding that this did not constitute an implied release of the right to a flow of water, Abbot J said: 'The owner [of a mill] is not bound to use the water in the same precise manner, or to apply it to the same mill; if he were, that would stop all improvements in machinery. If, indeed, the alterations made from time prejudice the right of the lower mill, the case would be different.'

Moore v Rawson (1824)

The plaintiff pulled down a wall which had contained some windows which enjoyed a right of light. He rebuilt it as a blank wall. Fourteen years later the defendant built a building in front of the blank wall. Three years after that the plaintiff put a window in the blank wall (where one of the original windows had been) and sued the defendant for obstruction. The Court held that the plaintiff had abandoned his right, and Littledale J suggested that two or three years' non-use of such a continuous easement would normally be sufficient to infer an intention to abandon permanently.

Tapling v Jones (1865)

The plaintiff had altered the size and location of his lower windows, so that the new windows could only partly be said to be in the same apertures as the ancient lights. He had also added new upper windows in a position which meant that the adjoining owner could not obstruct them without obstructing the ancient part of the lower windows. Subsequently the plaintiff restored the lower windows to their original size and position, blocked the upper windows and demanded that the defendant pull down his wall. The House of Lords held that (1) the plaintiff had done nothing unlawful in opening the new windows, and (2) by building the wall in a way which obscured the ancient lights the defendant's act was unlawful from the outset. It is important to note that it was only in so far as he was obstructing the original window apertures that the defendant was acting unlawfully.

Ecclesiastical Commissioners v Kino (1880)

A church was demolished with the intention of selling the site for development thereafter. After the church had been demolished, and before it could be sold, the owner of adjoining land started erecting a building on it that would have obstructed the light to the original church windows. The plaintiff successfully sought an injunction to restrain the adjoining owner from obstructing the windows of any building to be erected on the site of the church, in so far as such windows occupied the same position as those of the demolished church.

3.2.2 Implied release of discontinuous easements

Discontinuous easements are those which by their nature are not in use 24 hours a day. Simply not using such an easement, no matter how long that cessation of use continues, is not sufficient for the court to infer abandonment of the right.

Whether or not abandonment/implied release can be inferred therefore depends on a careful analysis of whether:

- the dominant owner has done an act or acts beyond mere non-user which evince an intention to permanently abandon;
- the servient owner has done an act or acts acquiesced in or agreed to by the dominant owner which restrict or wholly prevent exercise of the right; and/or
- the servient owner has acted to his detriment in reliance upon his reasonable assumption that the dominant owner intended to abandon.

In every case it is important to remember that it is primarily the intention of the dominant owner with which the court is concerned.

The interruption of enjoyment of an easement pursuant to a statutory obligation on the part of the servient owner will not effect an implied extinguishment of that right.

Ward v Ward (1852)

A right of way was not lost by non-user for a period of well over 20 years. Alderson B. held that: 'The presumption of abandonment cannot be made from the mere fact of non-user. There must be other circumstances in the case to raise that presumption. The right is acquired by adverse enjoyment. The non-user, therefore, must be the consequence of something which is adverse to the user.'

Cook v Corporation of Bath (1868)

The plaintiff had originally had a right of way through a back door, which he then closed off for 30 years. He then re-opened it, and had used it for four years before the defendant obstructed it. The Court held that there had been no abandonment, but noted that if the defendant had commenced building before the back door had been re-opened, the plaintiff would have been estopped from denying abandonment.

Gotobed v Pridmore (1971)

Buckley LJ set out the now standard test as: 'To establish abandonment of an easement the conduct of the dominant owner must, in our judgment, have been such as to make it clear that he had at the relevant time a firm intention that neither he nor any successor in title of his should thereafter make use of the easement...Abandonment is not, we think, to be lightly inferred. Owners of property do not normally wish to divest themselves of it unless it is to their advantage to do so, notwithstanding that they may have no present use for it.'

Williams v Usherwood (1983)

Two houses were built with a shared driveway. Ownership of the driveway was split, and each of the owners given rights of way over that part of the driveway in the other's ownership. Because of the layout of the houses, the driveway was only of practical use to one owner, who (in 1935) erected a fence separating the entire driveway from the other house. The owners of the dominant tenement never exercised the right of way as such. Ownership of the dominant tenement changed in 1937 and both owners thereafter treated the fence as the boundary between the properties. The Court of Appeal held that in those circumstances the defendant had acquired adverse possession of that part of the driveway not already owned by him, and that the plaintiff's predecessor in title had abandoned the right of way in about 1937 when she evinced an intention only to use an alternative driveway for access to her property.

Benn v Hardinge (1993)

Non-use of an implied right for the laying out of a way for 175 years did not raise any presumption of abandonment in circumstances where nobody had had any occasion to use the way in that period, there being an alternative means of access.

Bosomworth v Faber (1995)

A party enjoyed a right to a supply of water from a tank. He then entered into a licence entitling him to construct a new tank in a different location and take his supply from that. The licence was subsequently determined. The Court of Appeal held that the previous right was impliedly abandoned when the licence was accepted, the new tank installed, and the previous one demolished.

Carder v Davies (1998)

The plaintiff enjoyed a right of way along a roadway adjoining the dominant tenement. A wall had been built some 30 years earlier which meant that access could only be gained to the roadway part way along it. The plaintiff demolished part of the wall to allow access at a different point for her car. The Court of Appeal held that there had been no abandonment, Peter Gibson LJ saying: 'It is possible to abandon the use of part of a roadway. It is possible to abandon the right to use the easement in a particular manner. But it is a novel proposition to me, at any rate, that the easement owner with a completely unfettered right to enter the roadway at any point on the boundary of the owner's property, should limit that right as a matter of law so as to be confined to a particular access point, merely because of the existence of a wall or fence or hedge over or through which, while it stands, he would not conveniently be gaining access to the roadway.'

Jones v Cleanthi (2007)

The tenant of a flat under a long lease had a right to access and use communal bins at the rear of the building. In 1995 the landlord built a wall blocking such access pursuant to a statutory notice served by the local authority. The court held that this did not have the effect of extinguishing the claimant's rights once and for all, but that the landlord was not liable for interference with those rights since he was acting pursuant to a statutory obligation.

3.3 EXTINGUISHMENT BY STATUTE

Where lands are compulsorily acquired pursuant to statute, easements can be extinguished or suspended, depending on the particular statutory scheme in question. Generally speaking, a person having the benefit of an easement over the compulsorily acquired land cannot usually bring an action for disturbance of the easement. However, compensation for loss of the easement will generally be payable. There are too many different schemes to list here, and readers should turn first to the statute under which the land was acquired to determine the effect of any particular scheme.

3.4 EXTINGUISHMENT BY OPERATION OF LAW

There are also limited circumstances where easements can be extinguished by operation of law.

The most important of these is 'unity of possession', i.e. where the freeholds of the dominant and servient tenements, having been in separate ownership, come into common ownership. In such circumstances the easement which previously existed merges into the general rights of the property, in the same way that a lease merges with the freehold of a property when a lessee buys the freehold reversion. No such merger takes place where the adjoining properties are held in different capacities, for example where the same person owns the freehold of one plot and a long leasehold interest in the adjoining plot.

Kilgour v Gaddes (1904)

Two tenants of the same landlord fell out over the right of one to draw water from a well on the other tenant's premises. The tenant wanting to draw water claimed a right to do so by way of prescription, having used the well for the previous 40 years. The Court dismissed the claim for prescription. It held that, prescriptive rights being acquired by tenants for the benefit of the freehold reversion, and the freehold of both properties being held by the same landlord, the landlord could not acquire a right against himself.

Wall v Collins (2007)

The leasehold owner of a house under a 999-year lease was granted a right of way over a passageway. Several years later the freehold and leasehold ownership of the house merged, and the lease ceased to exist. The Court of Appeal held that the right of way did not cease to exist upon merger of the freehold and leasehold. Note here that the freehold and leasehold merged were both of the dominant land. The effect in this case was that the freehold owner of the dominant land continued to benefit from the easement (for the balance of the 999 years which would have been the term of the lease).

Apart from unity of possession, there are other, highly unusual circumstances which can give rise to extinguishment by operation of law. They are:

- where the necessity giving rise to an easement of necessity is removed;
- where the purpose for which an easement was created has come to an end; and
- by 'frustration', where circumstances have changed since the grant of the easement which mean that there is no practical possibility of the easement ever benefiting the dominant tenement again in the manner contemplated by the grant.

4 Common easement issues

4.1 INTRODUCTION

The following section deals with some specific problems and issues which you may well meet in relation to easements, namely:

- Alterations in the extent or mode of user.
- Alterations in the route of a right.
- Access for repairs and liability to repair.

4.2 ALTERATIONS IN THE EXTENT OR MODE OF USER

In the case of an express grant of an easement, the extent of user permitted is determined by the wording of the grant. In the case of an easement created by prescription, the extent of user permitted will be limited and defined by the use proved and which gave rise to the right. In both cases the Court will also have regard to the physical layout of the premises at the date of grant, since the extent of the user may be restricted by that. Consequently a substantial increase in the quantity of user may well be lawful in the case of an express grant, but is unlikely to be so in the case of an implied grant or easement created by prescription.

Subject to any qualifying words in the grant, an express grant will have the effect of authorising a mode or quality of user as general as the physical capacity of the land will allow. Therefore, where there is an express grant, changes in the mode of user will be lawful where they fall within the ambit of the express grant, even if they were not considered at the time of grant. However, where the grant relied upon is an implied grant, or prescription, changes to the mode of user will generally be unlawful.

South Metropolitan Cemetery Co. v Eden (1855)

In this case Jervis C.J. gave the following useful illustration: 'If I grant a man a way to a cottage which consists of one room, I know the extent of the liberty I grant; and my grant would not justify the grantee in claiming to use the way to gain access to a town he might build at the extremity of it. Here the grant is general, to use the road for the purpose of going to or returning from the land conveyed, or any part thereof; it is not defined, as in the case referred to.' While the wording of the grant or the situation of the land may impose an express or implied limit on the extent of use permitted, the right will generally be considered to be unlimited in extent.

Crossley & Sons Ltd v Lightowler (1867)

A riparian owner had acquired a prescriptive right to pollute water. The Court held that the dominant owner was not entitled to increase the pollution to the prejudice of others.

Wood v Saunders (1875)

The plaintiff was entitled by express grant to a right to drain water and soil from his mansion to a cesspool on the defendant's land. He extended his mansion. which increased the quantity of soil, and the defendant then stopped up the drain. The Court granted an injunction restraining the defendant from stopping up the drains, but only on the basis that an injunction was also granted preventing the plaintiff from allowing the drainage from the additional buildings to drain into them.

Finch v Great Western Railway (1879)

This case set out the useful principle, since approved by the Court of Appeal, that 'where there is an express grant of a private right of way to a particular place, to the unrestricted use of which the grantee of the right of way is entitled, the grant is not restricted to access to the land for the purposes of which access would be required at the time of the grant'. In other words, if I have an unrestricted right of way to a plot of land, I am entitled to access it for a different and significantly more intensive use than I used it for at the date of grant.

Todrick v Western National Omnibus Co. (1934)

A right of way was granted over a road which was seven feet nine inches wide at its entrance and supported by a retaining wall. A right of way granted 'with or without vehicles' was held not to include a right of way for heavy vehicles, including motorbuses.

Graham v Philcox (1984)

The plaintiffs lived in the first floor of a coach house which enjoyed a right of way over adjoining land. The plaintiffs later acquired the ground floor of the coach house and occupied the whole of it as one residence. It was held that the alteration of the dominant tenement by its enlargement did not affect the existence of the right of way, because there was no evidence that the plaintiffs' use of it was or would be excessive.

4.3 ALTERATIONS TO THE ROUTE OF A RIGHT

The servient owner generally has no right to alter the route of an easement unless such a right is an express or implied term of the grant.

An example of such an express term would be a right 'to pass and repass with or without vehicles from and to the [dominant land] over such part of the [servient land] being not less than twelve feet in width as the [dominant owner] may from time to time designate for that purpose'.

Of course, there is nothing to prevent an alteration of the route of a right of way by agreement between the dominant and servient owners. If it is an express agreement, it should comply with the formalities required by section 2(1) of the *Law of Property (Miscellaneous Provisions) Act* 1989, in particular by being in writing and signed by both parties. While it may still be possible to enforce an agreement not complying with those formalities by way of estoppel, it is plainly more prudent to reduce such agreement to writing.

Parking rights often expressly reserve the right of the grantor to control, regulate or alter the parking arrangements.

However, if an obstruction is placed across a right of way by the servient owner, the dominant owner is entitled to deviate onto adjoining land owned by the servient owner to connect the two parts of the way on either side of the obstacle. This right ceases when the obstacle is removed.

Celsteel Ltd v Alton House Holdings (1986)

A landlord of a block of flats was entitled to impose a new one-way traffic flow system with designated exits and entrances, since this did not represent a substantial interference with the dominant owners' enjoyment of their rights, but rather enabled the car park to be used conveniently and safely by all those entitled to use it.

Greenwich Healthcare NHS Trust v London & Quadrant Housing (1998)

The servient owner realigned a right of way in a way which, the Court held, improved road safety, and achieved an object of substantial public and local importance. The dominant owner did not object at the time of the realignment, but later sought an injunction. In the exceptional circumstances of the case, the Court refused to grant the injunction requested.

4.4 ACCESS FOR REPAIRS AND LIABILITY TO REPAIR

This section deals with the common law rights of access for repairs. The *Access to Neighbouring Land Act* 1992 also introduces statutory rights which may create complementary or additional rights, and which are covered in Chapter 11 below.

The dominant owner is entitled to enter the servient land to carry out repairs or to alter the servient land in order to accommodate the right granted. Using drainage rights as an example, the dominant owner is entitled to clear or repair drains on the servient land, but must restore the surface of the servient land to its previous condition after completing those works.

Generally speaking, a dominant owner is not obliged to keep the subject of the easement in repair. However, there are limited circumstances where the dominant owner may in fact be forced to repair. If pipes the subject of a drainage easement leak water carried from the dominant land onto the servient land, this will constitute a trespass by the dominant owner, who, if he wants to continue to use the pipes, must repair them.

Grants of rights of way are commonly expressed to be subject to the dominant owner contributing a proportion of the cost of maintaining the way. Such obligations are initially enforceable by the parties to the grant under contract law, but once the land changes hands, this is no longer possible. Successors in title can, however, often enforce contributions provided that the dominant owner continues and intends to continue to use the way – this is on the principle that where a person is granted a right to a use which causes damage subject to compensating for that damage, he must pay the compensation if he exercises that right.

Unless bound to do so by statute or contract, a servient owner is generally under no obligation to execute any repairs to the subject matter of an easement. Consequently a covenant by the servient owner in a grant will be enforceable in contract, but will not be enforceable once the servient land is transferred away.

Jones v Pritchard (1908)

The party wall between two houses contained a chimney with several flues, and each owner had a right to allow smoke from his fireplace to pass into the flue connected with it. The plaintiff sought damages from the defendant because the flue serving the defendant's fire but on the plaintiff's property was in disrepair and leaking smoke. The Court held that there was no obligation on the part of the defendant dominant owner to repair the chimney of the servient tenement.

Owners of Strataplan 58754 v Anderson (1990)

The court held that a dominant owner was entitled to enter the servient land to instal lighting if it was reasonably necessary to enable the way to be used safely and conveniently.

5
Rights of way, parking and unloading

5.1 INTRODUCTION

Rights of way are the most common easement encountered in practice.

The right to park and/or load and unload are sometimes explicitly or implicitly included within rights of way, but also merit separate consideration, which they receive below.

This section covers some of the more common problems connected with these easements.

5.2 RIGHTS OF WAY CREATED BY EXPRESS GRANT

The extent of a right of way created by express grant will be determined by the wording of the grant. There are a number of points which must be considered when attempting to construe such an express grant:

- If a plan is attached to the conveyance, the court will have regard to the plan in determining the extent of the right of way, even if the plan is 'for identification purposes only'.
- The court may look at other contemporaneous documents (e.g. planning permissions or sale contracts) to resolve ambiguity.
- Regard may also be had to the physical characteristics of the land at the date of the grant as an aid to construction.
- If there is no clear answer despite the points mentioned above, the grant will be construed against the grantor, so that any ambiguity is resolved in favour of the grantee.

Soames-Forsythe Properties Limited v Tesco Stores Ltd (1991)

A supermarket had been granted 'full and free right of way on foot only' along pedestrian walkways leading from the supermarket to car parks. The Court held that in the circumstances of the grant, the parties must have contemplated that the customers would be supplied with supermarket trolleys, and the use of trolleys on the walkways was therefore not a trespass.

Charles v Beach (1993)

The servient owner granted a right to use 'the path or roadway' lying between two properties. At the date of the grant two-thirds of the 'path' had been a flower bed. The Court of Appeal held that the dominant owner was entitled to access over the whole of the 'path', the flower bed being a sufficiently transient insubstantial an obstacle that its existence could not be said to have limited the grant.

White v Richards (1994)

An express right of way provided for the 'right to pass on foot and with or without motorvehicles' over a track so far as necessary for the use and enjoyment of the dominant tenement. At the date of the conveyance, the track was 8′ 10″ wide and was formed of crushed stone covered with a well-worn layer of bitumen. The Court of Appeal held that the reference to 'motorvehicles' had to be construed in the light of the physical characteristics of the track, and was therefore restricted to vehicles with a wheelbase of no more than 8′ and a laden weight not exceeding 10 tons. Consequently the track's use by 14-16 heavy lorries per day was considered to be excessive use.

5.3 RIGHTS OF WAY CREATED BY PRESCRIPTION

A right of way created by prescription will be strictly delimited, by the actual use made of the way during the period relied on for the creation of the right.

It is important to note, however, that while a change in the nature of use (for example from agricultural to commercial) will not permit the acquisition of a right to the new use (unless that use continues for 20 years itself), an increase in the amount of use will be permitted.

RCP Holdings v Rogers (1953)

The defendant farmer owned a field and was able to establish that the field had been accessed along the way (over the plaintiff's golf course) for agricultural purposes for around 70 years. The defendant had recently established a campsite for caravans on the field and wanted to use the way as access for the caravans. The Court held that to do so would be a trespass, saying: 'the way here was a way limited to agricultural purposes, and that to extend it to the use proposed would be an unjustifiable increase of the burden on the servient tenement'.

Woodhouse & Co Ltd v Kirkland (Derby) Ltd (1970)

The plaintiff had established a prescriptive right of way over a passageway owned by the defendant. The passageway was used by the plaintiff's customers. Although the use by the plaintiff's customers had increased enormously since the right was established, the Court held that there had been no excessive user since the use was not of a different kind, nor for a different purpose.

Giles v County Building Constructors (Hertford) Ltd (1971)

The defendant developer had a right of way by prescription to two houses. He intended to demolish the houses and replace them with a block of six flats, a bungalow and seven garages. The Court held that since the development did not involve a radical change in the character or identity of the dominant tenement, the proposed use would not constitute an excessive user of the right of way.

Loder v Gaden (1999)

For many years a lane had been used for agricultural purposes. Just less than 20 years prior to the commencement of proceedings the dominant owner had begun using the lane for road haulage purposes, and used the lane increasingly intensively during that period. The Court held that although a right of way by prescription was established, it was limited to agricultural purposes, since there was nothing to indicate to the servient owner prior to the changed and increased usage that any right beyond use for agricultural purposes was intended.

5.4 RIGHT OF WAY TO DOMINANT TENEMENT ONLY

Generally speaking a right of way may only be used for gaining access to the dominant land identified as such in the grant. Exceptionally, however, if the use for the benefit of additional land (i.e. land not identified as the dominant land in the grant) is ancillary to or part and parcel of the use originally granted, it may be allowed.

Harris v Flower & Sons (1904)

The principle was set out clearly in this case as follows: 'If a right of way be granted for the enjoyment of close A, the grantee, because he owns or acquires close B, cannot use the way in substance for passing over close A to close B'.

Bracewell v Appleby (1975)

The defendant owned a house which enjoyed a right of way 'of the fullest description' to it. The defendant bought an adjoining plot of land and built another house on it. The Court held that he had no right to extend the grant of the easement to the adjoining plot.

Jobson v Record (1997)

A farm and farmhouse had a right of way over a road 'for all purposes connected with the use and enjoyment of the property hereby conveyed being used as agricultural land'. The dominant owner wanted to use the road to remove timber which had been felled on land adjoining the farm and stored on the farm. The Court held that this use was either (1) an attempt to use the road for the benefit of the adjoining plantation, or (2) that timber storage was not an agricultural use and so, in either case, was not permitted by the terms of the grant.

National Trust v White (1987)

The National Trust had the benefit of a right of way over a track 'for all purposes' to the historic site of Figsbury Ring. The local authority created a 1.5 acre car park at the side of the track (on land not forming part of Figsbury Ring) for the benefit of visitors. The Court held that using the track for access to the car park for the purpose of visiting Figsbury Ring was ancillary to the enjoyment of Figsbury Ring, and therefore within the terms of the grant. It should be noted, however, that if users of the car park were not intending to visit Figsbury Ring, such use would not have been within the grant.

Das v Linden Mews Ltd (2002)

The owners of two houses in a mews enjoyed rights of way to their houses. They also owned adjoining gardens beyond their houses which did not enjoy such a right but which they wanted to use to park their cars. The Court held that the parking of cars on the gardens was not ancillary to the use of the way for the purpose of grant. The decision would have been different if the garden had been part way along the right of way. This case is also notable because the Court of Appeal remitted the case to the trial judge for reconsideration as to whether damages should be awarded in lieu of an injunction, and gave strong indications that it thought they should.

5.5 USE OF WAY INTERFERING WITH THE RIGHTS OF OTHERS

Where more than one person enjoys a right of way, use by one person having that right to an extent which interferes with the enjoyment of the same right by others will be considered to be excessive, and will be restrained by injunction if necessary.

Rosling v Pinnegar (1986)

A narrow and winding lane provided access to a Georgian mansion and 25 dwellings in a village. All of these properties enjoyed a right of way in common with all other persons entitled thereto at all times and for all purposes with or without vehicles to use the lane. The owner of the mansion opened it to the public, and the villagers sought an injunction to stop people using the lane by public or general invitation. The Court held that the use of the lane by visitors to the mansion was excessive because it interfered unreasonably with the use of the lane by the villagers, and that it was appropriate to grant injunctions limiting the size of vehicles which could use the lane and the days upon which and times at which the owner could invite the public to use the lane.

5.6 OTHER MATTERS AFFECTING RIGHTS OF WAY

A right of way does include a right to 'vertical swing space', i.e. to have the right of way free from obstruction to such height as is reasonable given the extent of the grant. The reasonable height for a vehicular right of way would obviously be higher than that for a pedestrian-only right.

However, there is no right for the dominant owner to have horizontal swing space beyond the furthest extent of the way. Consequently the dominant owner has no cause for complaint if the owner of land abutting the right of way builds right up to the edge of the way.

5.7 PARKING AND UNLOADING

A right to park a car somewhere in a defined area is capable of subsisting as an easement. However, some care needs to be taken when assessing whether such a right has been obtained by prescription, since, if the Court considers the use to have been so substantial that all reasonable use of the affected land is prevented, it will find that the right exercised was not capable of being an easement.

This is perhaps best illustrated by considering a block of 20 flats with ten parking spaces outside. Where all flat owners have a right to park on any of these spaces, when available, this would amount to an easement. However, where ten of the owners have each been granted an exclusive right to use a different numbered space, this would probably constitute a lease or a licence instead.

A right to load and unload is implicit in a right to park, whether or not mentioned expressly.

Bulstrode v Lambert (1953)

A right to 'pass and repass' over a cul-de-sac for the purpose of obtaining access to an auction mart also implied a right to halt vehicles at the end of that cul-de-sac as long as was necessary for the purpose of loading and unloading. What amounted to an implied right to park for the purpose of loading and unloading was implied because such a right was necessary for the enjoyment of the express right to pass and repass.

Hair v Gillman (2000)

A right to park a car on a forecourt which was big enough to be able to accommodate two or three other cars was held to be capable of being an easement and fell on the easement side of the line between rights in the nature of an easement and rights in the nature of an exclusive right to possess or use.

Saeed v Plustrade Limited (2002)

The tenant of a flat was granted a right in common with others to park on such part of the forecourt as might from time to time be specified by the landlord as reserved for parking when space was available and subject to such regulations as the landlord might make from time to time. The landlord proposed to reduce 12 spaces to four spaces, and to charge for them. On appeal it was confirmed that the landlord was only entitled to change the location of the spaces, not to reduce the number of spaces, or to charge for them.

Moncrieff v Jamieson (2007)

The House of Lords decided in this Scottish case that a right to park could be an easement. A house and grounds upon which it was impossible because of its steepness and the existence of a wall, to park, claimed a right to park on the servient land. The Court accepted, on the particular facts of the case, that a right to park was ancillary to the right of way, and emphasised that the use of land for parking would have to be very extensive indeed if it were to be sufficient to defeat a claim of a right to park.

6
Rights of drainage and water

6.1 DRAINAGE

Drainage rights can be broken down as follows:

- the right to lay down or install pipes;
- the right to use pipes;
- the right to a supply of water or to discharge water/waste;

The right to repair is dealt with above. The same categories and principles apply equally to gas pipes, and to electricity of telecommunications cabling.

6.1.1 Right to lay pipes

The extent of such a right depends entirely upon the wording of the grant. Plainly, such a right cannot arise by prescription. Further, the courts have tended to construe the wording of such grants rather restrictively.

Simmons v Midford (1969)

A right was granted 'to lay and maintain drains, sewers, pipes and cables over under and along' a strip of land, and 'the free and uninterrupted passage and running of water, soil, gas and electricity there through and the right to enter upon and open up the said land for the purpose of laying, maintaining and repairing the said drains'. The owner of the servient land wanted to connect his drain into that laid and used by the dominant owner. The Court held that (1) the pipe belonged to the dominant owner, and/or (2) that the dominant owner was entitled to exclusive use of the pipe, and, in either case, the servient owner failed.

Trailfinders v Razuki (1988)

A lease reserved the right for 'the free and uninterrupted running and passage of water, soil, gas and electric current from other buildings and lands of the landlords...through the channels, sewers, drains, water courses, pipes, wires and other conduits which are now, or may hereafter during the term hereby granted be in, under or over the demised premises'. The plaintiffs wanted to lay computer cables over the defendant's land. The judge held that although the plaintiffs could enter the defendant's land to maintain or replace existing cables, they were not permitted to lay new cables.

Martin v Childs (2002)

A house enjoyed 'a right to run water, electricity and other services through any pipe, cables, wires or other channels ('the Conduits')...and the right to enter onto...the Retained Land [of the vendor] for the purpose of installing, repairing, maintaining, cleansing and inspecting of the Conduits'. The dominant owner entered the servient land to install a new pipe in a different position in order to secure an adequate water supply. The Court of Appeal held that the grant only allowed the installation of conduits over the same routes as those existing at the date of grant, and did not permit the size or position of existing water pipes to be altered; 'installing' referred to the provision of new services which were not there at the date of the grant.

6.1.2 Right to use/right to supply

Beauchamp v Frome Rural District Council (1938)

The plaintiff's farm enjoyed 'the right as now enjoyed in common with others having the same right to a water supply to the hereditaments hereinbefore described through pipes from a spring'. The local authority owned the spring and the pipes, and laid new pipes which reduced the supply to the farm. The Court held that while the farmer was only entitled to the residue of the water after the other persons entitled had taken what they wanted, this only applied to installations in place at the date of the grant. Consequently the farmer was entitled to object to the new and larger service pipes.

Duffy v Lamb (1997)

The dominant owner had a right to the passage of electricity over the servient land. The servient owner switched off the dominant owner's electricity sub-meter, which was on the servient land. The Court held that the servient owner was liable to the dominant owner. The servient owner's obligation was held to have been to take no positive step to prevent the entry of electricity onto his land, or its subsequent passage through it.

6.2 WATER RIGHTS

Rights in relation to water require a slightly different approach to other rights, since the owner of land which contains or abuts a watercourse enjoys rights at common law closely analogous to easements (commonly called 'riparian' rights). The rights are analogous to easements because they involve the riparian owner having rights and obligations affecting land other than the land he owns. When enjoyed by the owner of the freehold, such rights are fundamentally part of the fee simple.

However, such rights can also be acquired by grant, reservation or prescription, and are thus severable from the freehold.

It is crucial to note that common law riparian rights have been substantially limited by statute, in particular the *Water Resources Act* 1991.

Portsmouth Borough Waterworks v London Brighton and South Coast Railway (1909)

In this case Parker J said:

> 'When a riparian owner sells part of his estate, including land on the banks of a natural stream, it is not necessary to make any express provision as to the grant or reservation of the ordinary rights of a riparian proprietor. These rights are not easements to be granted or reserved as appurtenant to what is respectively sold or retained, but are parts of the fee simple and inheritance of the land sold or retained. If it be desired to alter or modify these rights, it can only be done by the grant or reservation of such rights in the nature of easements as the nature of the case

> may require. If no such rights are granted or reserved, the vendor remains, and the purchaser becomes, a riparian owner, and retains or acquires all the ordinary rights of a riparian owner.'

A riparian owner has three naturally occurring rights in relation to the watercourse:

- He can use the water for certain purposes connected with his riparian land.
- He has a right for the water to flow unobstructed onto and off his land.
- He has a right for the water not to be polluted.

6.2.1 Water flowing through a natural channel

Riparian rights apply to surface water flowing through a natural channel, i.e. a river or stream. They apply equally to water flowing underground through a known and physically defined channel.

McCartney v Londonderry and Lough Swilly Railway Co (1904)

A railway line crossed a natural stream, so that the railway company owned a section of land about 8 feet wide abutting the stream. They inserted a pipe which carried water half a mile from the stream, along the line of the railway line, to other land they owned. Here it was used for the company's locomotives. The lower riparian owner stopped up the pipe and was sued by the railway company. The Court held that the lower riparian owner was entitled to stop up the pipe, saying:

> 'There are, as it seems to me, three ways in which a person whose lands are intersected or bounded by a running stream may use the water to which the situation of his property gives him access. He may use it for ordinary or primary purposes, for domestic purposes, and the wants of his cattle. He may use it also for some other purposes – sometimes called extraordinary or secondary purposes – provided those purposes are connected with or incident to his land, and provided that certain conditions are complied with. Then he may possibly take advantage of

his position to use the water for purposes foreign to or unconnected with his riparian tenement. His rights in the first two cases are not quite the same. In the third case he has no right at all.'

Rugby Joint Water Board v Walters (1967)

The defendant was taking water from the River Avon in order to irrigate his land. Occasionally he took as much as 60,000 gallons a day. The evidence was that although this had no visible or measurable effect on the river, it amounted to a considerable volume of water, only a very small part of which was returned to the river. The Court held that the plaintiff was entitled to an injunction to prevent the abstraction on the basis that a riparian owner is not entitled to take water from a stream for extraordinary purposes without returning it to the stream in substantially the same quantity.

6.2.2 Surface water not in defined channel

Where surface water comes from springy or boggy ground and flows in no particular direction, or occasionally surfaces at one spot, the owner of the land has a right to dispose of the water by draining the land or in any other way he wants. This rule applies even where the water would eventually have reached the course of a natural stream.

Broadbent v Ramsbotham (1856)

A pond, when it occasionally exceeded a certain depth, overflowed and spread over the surface of the land, some of the escaping water gradually making its way to and augmenting a natural stream. The owner of the pond was held not liable for draining the pond.

6.2.3 Statutory restrictions on riparian rights

Under section 24(1) of the *Water Resources Act* 1991, no person shall abstract water or cause or permit any other person so to abstract any water, except in pursuance of a licence under the Act granted by the Environment Agency, and in accordance with the provisions of that licence. The general rule is that a licence is required to extract water from a 'source of supply', which is defined to mean any inland waters except 'lakes, ponds or reservoirs which do not discharge to other inland water or underground strata...in which water is or may at any time be contained'.

Only occupiers of land contiguous to the source of supply, or a person having a right of access to such land, are entitled to apply for licences. The Environment Agency has a discretion, subject to the factors it is required to take into account by the Act, as to whether or not to grant a licence.

The licence, if granted, must specify whether it is to remain in force until revoked or is to expire at a specified time, how much water may be extracted during a specified period or periods, how to determine what quantity of water is to be taken to have been extracted, and the purpose for which the abstracted water is to be used.

There are some cases where a licence is not required. The most important of these in practice is that a licence is not required for any abstraction for domestic purposes and agricultural purposes other than spray irrigation unless the quantity abstracted exceeds 20 cubic metres in any period of 24 hours.

6.2.4 Prescriptive rights in natural watercourses

If water flows through a natural watercourse with a defined channel, it is possible for a person to acquire rights by prescription which interfere with what would otherwise be the natural rights of the riparian owners above and below the dominant land.

Cooper v Barber (1810)

A riparian owner through whose property a man-made feeder from the River Lavant flowed, was held to have acquired a prescriptive right to pen back a stream for the purposes of irrigating his land, but not so as to cause flooding to his neighbour's land.

Roberts v Fellows (1906)

A lower riparian owner was held to have acquired a prescriptive right to go on the land of an upper riparian owner and bank up the stream.

6.2.5 Prescriptive rights in artificial watercourses

Rights can equally be acquired by prescription in permanent artificial watercourses, as if they were natural watercourses.

However, where a temporary artificial watercourse is concerned it is not possible to acquire a right by prescription.

Arkwright v Gell (1839)

The lower riparian owners were occupiers of mills located on a stream which drained a mine. They attempted to compel the mine owners to continue such discharge. Giving judgment, Lord Abinger said: ' The stream...was an artificial watercourse, and the sole object for which it was made, was to get rid of a nuisance to the mines, and to enable their proprietors to get the ores which lay within the mineral field drained by it: and the flow of water through that channel was, from the very nature of the case, of a temporary character, having its continuance only whilst the convenience of the mine-owners required it, and, in the ordinary course it would, most probably, cease when the mineral ore above its level should have been exhausted'. Such a temporary watercourse could not, the Court said, give rise to a permanent right.

6.2.6 Purity of water/pollution

Wright v Williams (1836)

The right to discharge polluted water into another's watercourse was acquired under the *Prescription Act* 1832. However, the *Water Resources Act* 1991 now provides that a 'person commits an offence if he causes or knowingly permits any poisonous, noxious or polluting matter or any solid waste to enter any controlled waters, unless the entry occurs or the discharge is made under and in accordance with a consent under [the Act].' Consequently, unless a discharge is made pursuant to the requisite consent, it cannot give rise to a prescriptive right.

Young (John) & Co v Bankier Distillery Co. (1893)

A mine had been discharging pit water into a stream, which polluted it to such an extent that it was unfit for the production of whisky. The Court held that a riparian owner on the banks of a natural stream is entitled as a natural right to have the stream flow past his land without sensible alteration to its character or quality, and that the distillery was therefore entitled to an injunction preventing the discharge.

7 Rights of support and protection

7.1 INTRODUCTION

The support and protection of land and buildings is an area into which the general law of tort has made considerable inroads. Consequently, it is not enough simply to consider these matters from a property law perspective, and at the end of this chapter there are a number of cases which have been brought before the courts as claims in tort.

7.2 SUPPORT

When one thinks about 'support' in the property context, one instinctively focuses on the simple situation of a building and the land upon which it stands. Whilst buildings do of course require support, in order to understand the law relating to support it is better to start even more fundamentally than this and to remember that even an undeveloped piece of land requires support – it will usually be supported vertically by subjacent soil strata and, ultimately, by bedrock. It will be supported laterally by the adjacent land – dig a hole in soil to any depth and the edges of the hole will soon 'fall-in' if not supported, illustrating the need for lateral support!

The position quickly becomes more complicated when one starts to consider that soil has different properties according to its type, and also that it has the ability to retain and release liquids, primarily water. All of these matters can affect the ability of soil to provide support.

Moreover, introduce a load onto the land in question, for example by placing a building upon it, and, in mechanical terms, this will result in a downward vertical pressure on that land and the subsoil, with the land supplying an equal and opposite vertical reaction to prevent the building from sinking. However, the weight of the building will also generate additional lateral forces within the soil beneath it – the soil immediately underneath the building will have an increased tendency to move out but will be restrained laterally by the adjacent soil. In a simple case, the greater the weight of the building, the greater the lateral forces induced in the soil and the more important it becomes that the adjacent soil is able to provide an equal and opposite reactionary force.

In practice this means that whilst excavating to a certain depth, X, need not cause subsidence in adjacent land in its natural state, if that adjacent land happens to be loaded with a building, it may well subside.

Whilst the foregoing is a gross over-simplification of the science of soil mechanics, it will nevertheless assist in understanding some of what follows.

7.2.1 Natural support to land (no buildings)

A landowner is prima facie entitled to support for his land from the adjacent land, and, if he does not own it, the subjacent land. This is not a matter of the law relating to easements, but is simply an incident of the ownership of the land.

Dalton v Angus (1880-81)

In this case (the facts of which appear below in paragraph 7.2.2) Lord Blackburne observed that the owner of land had a right to support from the adjoining soil. This is not a right to have the adjoining soil remain in its natural state, but a right to have the benefit of support; that right is infringed as soon as, but not until, damage is sustained in consequence of the withdrawal of the support.

However, the obligation of lateral support only extends so far, namely as far as so much of the adjoining land as in its *natural state* is required to support the claimant's land.

Corporation of Birmingham v Allen (1877)

The claimants owned a gas works and the defendants owned a colliery. The two pieces of land were separated laterally by an intervening piece of land which had been worked for coal and under which, consequently, there was a cavity. The effect of this cavity was that when the defendants mined their land, subsidence resulted to the land belonging to the claimants situated beyond the intervening land. The claimants sought an injunction to restrain the defendants from continuing to mine. It was common ground that if the intermediate land had been in its natural state then the defendants' operations would have cause not damage to the claimants land. It was held that the entitlement to support was an entitlement to support from the 'neighbouring landowner', which in this context meant the owner of that portion of land which in its natural state was necessary for the support of the land in question, and nothing beyond that. The intervening land was the neighbouring land – the only reason that it did not in fact provide the required support was because it was no longer in its natural state. The claimants had no entitlement to support from land further away than this and were not granted the injunction they sought.

As mentioned earlier, the liquid content of soil may contribute to its load-bearing properties. What is the position when an adjoining owner drains his land of water, causing the water under his neighbour's land to migrate, thereby reducing its load-bearing ability? The natural right of support mentioned above does not extend so far as to prevent an adjoining land-owner from doing this (but see also below *Lotus Ltd v British Soda Ltd*) – at least in so far as percolating water is involved. This is consistent with the law that no one has a proprietary right to water percolating beneath his land.

Popplewell v Hodkinson (1868-69)

The defendant builder, who had been engaged to construct a church, excavated deeply to firm soil for the purposes of constructing the foundations. During this process the wet and spongy adjacent land upon which the claimant's cottages were built was drained and its surface subsided. The claimant sued for damages. The Court held that as a matter of principle an adjoining owner was at liberty to drain his own land, although the result of his doing so was to cause subsidence to his neighbour's land. Accordingly, the claimant's claim failed. The Court did however note that the position might be different where A had granted land to B for some special purpose like building, whilst retaining some adjoining land. In such circumstances A's rights to use his land might, in accordance with the maxim that a man cannot derogate from his own grant, be restricted to acts which did not make the land granted to B less fit for the special purpose for which it was granted.

It appears that the decision in *Popplewell v Hodkinson* only applies in the case of the migration of water – in other cases involving quicksand and pitch (see below), the Courts have upheld a natural right of support.

Jordeson v Sutton, Southcoates & Drypool Gas Co. (1898)

This case illustrates the difficulties in drawing a distinction between percolating water, and other liquids. The defendants carried out works of excavation on their own land, and in doing so cut through a stratum of quicksand, known as 'running silt', lying beneath their land and also beneath the claimant's land. The consequence was that the claimant's adjoining land subsided. The Court of Appeal was divided as to whether, as a matter of fact, based upon the evidence, the claimant's house was supported by a stratum of water, or by a bed of silt. The majority held that it was supported by a layer of silt, as distinct from water. It followed that the decision in *Popplewell v Hodkinson* was inapplicable, and the Court considered that the right to support had been infringed by the defendants.

Trinidad Asphalt Co. v Ambard (1899)

The respondents dug out their own land to exploit asphalt deposits lying underneath it. A consequence of this was that asphalt lying under the adjacent land belonging to the appellants oozed into the respondents' land, causing subsidence to the appellants' land. The Privy Council considered that asphaltum was a mineral rather than water, and the appellants were granted an injunction and damages.

It should also be noted that whilst there is nothing in principle to prevent a landowner from draining his land of water even when this will result in the removal of water from the substrate of adjacent land, where the movement of this subterranean water causes the dissolution of minerals supporting the adjacent land, the pumping out of the resultant liquid may infringe the right to natural support.

Lotus Ltd v British Soda Ltd (1972)

The claimants' and the defendants' lands were supported by subterranean salt beds. Serious damage was caused to the claimants' land by an activity known as wild brine pumping carried on by the defendants on their land. The defendants were salt producers and extracted brine (water saturated with salt) from boreholes on their land. As the brine was pumped from the borehole, water was drawn through the salt beds under the adjacent land towards the borehole. This flow of water caused the dissolution of the salt beds under the claimants' land, the resultant brine flowing to the defendants' borehole and being extracted. As the salt beds below the claimants' land dissolved, the land subsided. The Court held that the claimants had a right to have their surface land supported by the subjacent strata of minerals and, referring to the case of *Jordeson v Sutton, Southcoates & Drypool Gas Co.* (above), that there was no significant difference in principle between removal of support such as wet sand and an operation which consists first in causing a solid support to liquefy and then removing the resulting liquid. Accordingly, the claimants were award damages and granted an injunction.

Although land enjoys no natural right to support from percolating water, can a right to such support be acquired through long user (see Chapter 2)? Comments in *Jordeson v Sutton, Southcoates & Drypool Gas Co.* suggest that this might be possible, but in the case of *Brace v South East Housing Association Ltd* a differently constituted Court of Appeal proceeded upon the assumption that it is not.

7.2.2 Buildings supported by land

In contrast to the position with land, there is no natural right to the support of a building. Any such right must either be acquired as an easement (see generally Chapter 2), or arise indirectly as a result of the law of tort, or in some cases, as a result of the *Party Wall etc. Act* 1996.

However, it is important to note the following subtlety: although there is no natural right of support for a building, if damage is caused to a building as a result of an interference with *the natural right of support* of the land upon which the building sits, so long as the subsidence is not the result of the additional weight of the building upon the land, the owner of the land can claim damages not just for the subsidence damage to the land, but also for the consequential damage to the building.

Brown v Robbins (1859)

The claimant was the owner of a house built on solid ground, adjacent to which was a garden belonging to another person from under which certain minerals had been extracted. On the other side of the garden lay land belonging to the defendant. The defendant extracted minerals from that part of his land adjoining the garden which in turn adjoined the claimant's house. As a result, the claimant's land sank and his house was damaged. It was held that inasmuch as the sinking of the claimant's land was in no way caused by the weight of the house, the claimant was entitled to damages for the damage to the house, even though he had not acquired any right of support for the house itself independent of the natural right of support for the land upon which it was built.

The fact that a building may acquire a right of support in its own right by long user is established by the case of *Dalton v Angus*.

Dalton v Angus (1881)

This case concerned adjoining properties separated by a wall. The properties were over 100 years old, but about 27 years before the incident in question, the claimant had converted his property, which had been a dwelling, into a coach factory. The consequence of this was to substantially increase the pressure on his land and laterally on the land beneath the adjoining property of the defendants. This was done openly but without any express consent on the part of the defendants. The defendants had recently demolished their property with a view to developing their site. This did not cause any problems. However, they then went on to excavate on their property for the purpose of constructing a cellar, and in so doing they dug lower than the foundations of the claimant's coach factory. The consequence of this was that the coach factory collapsed and the claimant brought an action on the basis that its coach factory, having been constructed openly more than 20 years ago, had by long user acquired a right to the support of the adjoining land. The case was considered by the House of Lords who held in favour of the claimant. The reasoning of their Lordships' differs, but it is generally accepted that this case establishes that a building can acquire, either through common law, the doctrine of lost modern grant or under the *Prescription Act* 1832, a right to support from adjacent, and indeed subjacent, land. It also follows from this case that a right of support can be acquired other than through long user – for example see paragraphs 2.2 and 2.3.

7.2.3 Buildings supported by buildings

As in the case of buildings supported by land, there is no natural right to have your building supported by your neighbour's building; any such right must either be granted, acquired or arise indirectly by reason of the law of tort. Acquisition may, again as in the case of buildings supported by land, be by long user or another method referred to above at paragraphs 2.2 and 2.3.

The reader is also referred to the *Party Wall etc. Act* 1996 (see paragraph 11.1) which in many cases will regulate work to adjoining structures and in some instances confers, subject to certain safeguards, greater rights to carry out work than would otherwise exist at common law.

7.2.4 When is there an infringement of right of support?

When considering whether or not there has been an infringement of a right of support, there are some subtle distinctions between situations involving the right to support from land and the right to support from a building.

In the case of the right to support from land (whether the support exists for the benefit of other land and/or buildings on other land), the right is *not* to have the adjacent (or subjacent) land preserved as it is, but *is* a right not to have the support enjoyed by the dominant tenement adversely affected by work carried out to the adjoining land. However, the right of support does not impose upon the adjoining owner any positive obligation to take active steps to maintain the support (but see paragraph 7.4 below).

Sack v Jones (1925)

In this case the plaintiff and the defendant owned adjoining houses separated by a party wall. Each house had a mutual right of support from the other. It was alleged by the plaintiffs that, due to lack of repair and underpinning at the defendant's house, it was subsiding, dragging the party wall over with it, damaging the plaintiff's house. The Court was not satisfied on the evidence that this was so, but said that even if the evidence had substantiated the plaintiff's claims, she would not have succeeded because although the defendant's house was subject to an easement of support in favour of the plaintiff's house, the defendant was under no obligation to keep her own house in repair for that purpose.

Furthermore, in the context of rights of support (but again see paragraph 7.4 below), a landowner is not responsible for the effects of nature, even though the effects occur only because of his use of his land, if he has used it in a normal manner.

Rouse v Gravelworks Ltd (1940)

The plaintiff farmer owned a field beside a gravel pit belonging to the defendants. The defendants, in the course of their business as gravel merchants, removed gravel from their land at an area bordering the plaintiff's field. The excavated area filled with water by rain and percolation, forming a pond. The water, by erosion and by being blown against the plaintiff's field, caused some removal of material from the plaintiff's field, and prevented the plaintiff from making profitable use of a narrow strip along the boundary. The court held that the plaintiff had no cause of action, as the defendants were entitled to remove gravel from their land and the damage to the plaintiff's land arose not from the direct actions of the defendants (i.e. the removal of the gravel) but by the natural agencies of rain, percolation and wind, over which the defendants had no control and so they were not responsible.

In the case of a right to support from a building, the position was explained in *Bond v Nottingham Corporation*, apparently in the context of a right to support acquired by implication or long user.

Bond v Nottingham Corporation (1940)

The plaintiff owned a property adjoining a property which the defendant corporation proposed to demolish. The plaintiff's property enjoyed a right of support as against the property to be demolished. The plaintiff argued that the defendant could not demolish the adjoining property without providing alternative mean of support for his own property. The Court agreed and explained the nature right of support, saying that:

- A servient owner is under no obligation to repair that part of his property which provides the support for his neighbour's; if he leaves it and it falls into decay compromising the support, the neighbour cannot complain.
- However, the neighbour need not sit by and watch the deterioration of the support; he is entitled to enter and carry out the necessary repairs to ensure that the support continues.
- What the servient owner is not entitled to do is by an act of his own to remove the support without providing an equivalent.

The above cases must be approached with some care, having regard to the development of a tortious duty of care (see below, at paragraph 7.4).

In the unusual situation in which a right to support arises by way of express grant, then it will be necessary to look at the wording of the grant to determine the precise scope of the right.

An infringement of a right to support may be the subject of a claim for damages, or in an appropriate case, an injunction. The cause of action arises when damage occurs. Where there are successive subsidences, a new cause of action arises on each occasion. Where support has been compromised, but no physical damage has yet been caused, but may be caused in the future, it seems that it is not possible at common law to claim damages for the loss in value of the affected property, although it is thought that in appropriate circumstances it might be possible to obtain damages in equity in lieu of an injunction.

West Leigh Colliery Co. Ltd v Tunnicliffe & Hampson Ltd (1908)

The plaintiffs sought damages for subsidence caused to their land by the extraction of minerals from beneath their land and adjoining land. The risk of further subsidence at some point in the future decreased the value of the land. The House of Lords held that in assessing the proper amount of damages, no account should be taken of the depreciation in the value of the plaintiffs' property brought about by the apprehension of future damage, because this gave no cause of action by itself.

Hooper v Rogers (1975)

In this case, the plaintiff's farmhouse stood on a slope. The plaintiff and the defendant were owners in common of the immediately surrounding land, which sloped steeply away from the plaintiff's farmhouse. The defendant used a bulldozer which deepened a track which cut across the slope. In due course the result would be the collapse of the plaintiff's farmhouse. He sought from the court damages in lieu of an injunction requiring the defendant to reinstate the natural angle of the slope. On appeal the Court held that the judge had had jurisdiction to grant a mandatory injunction to require works to avoid future damage, and accordingly had jurisdiction to grant damages in lieu of such an injunction.

7.3 PROTECTION

The starting point is that the law does not recognise an easement entitling the dominant tenement to be protected from the weather.

Phipps v Pears (1965)

In this case two houses adjoined in that their flank walls were up against one another but not bonded together. The defendant demolished his house, exposing the flank wall of the plaintiff's house to the elements. That flank wall had never been rendered, and rain penetrated the wall, froze, and caused cracking. The plaintiff claimed damages from the defendant. His claim was dismissed and this decision was upheld on appeal by Lord Denning, who said held that an easement of protection from the weather was unknown to the law and that everyman is entitled to pull down his house and that if it exposes another house to the weather, then that is the problem of the owner of the other house.

The observation to the effect that a man may pull down his house and expose his neighbour's to the weather with impunity must now be qualified by reference to the duties which it is now accepted are imposed by the law of tort on a landowner (see below at paragraph 7.4, and especially *Rees v Skerrett*)

7.4 DUTIES IMPOSED BY THE LAW OF TORT

As has been observed above, the law of easements and the natural right to support may leave a landowner vulnerable to his adjoining owner's activity (see above *Rouse v Gravelworks Ltd*) or indeed inactivity (see above *Bond v Nottingham Corporation*). In the second half of the last century the law developed so as to impose upon a landowner in certain circumstances a positive duty towards a neighbour. The precise extent of the duty, or what is required to discharge the duty, is not the same as the more familiar duty of care arising in negligence, and is to some extent dependent upon the characteristics and capabilities of the person upon whom it is imposed. The cases below illustrate the legal development of this duty.

Goldman v Hargrave (1967)

A tree on the defendant's land was set alight by lightning. He chopped the tree down and, although it would have been a simple matter to spray the tree with water to extinguish the remains of the fire, instead left it to burn out. The weather changed, with the result that the fire spread over his land and caused damage on the land of the plaintiff. The Privy Council held that even though the hazard was an Act of God, rather than the result of anything which the defendant had done, the defendant nevertheless owed a duty to his neighbour. In determining the scope of such a duty the law had to have regard to the fact that the occupier has had the hazard thrust upon him, and his interest in the land or his resources may be very modest either in relation to the magnitude of the hazard or compared with those of his neighbour. A rule which required, in the interest of his neighbour, a physical effort of which he was not capable, or which involved him in excessive expenditure would be unenforceable and unjust. Instead, in general terms, the existence of the duty must be based upon knowledge of the hazard, ability to foresee the consequences of not acting, and the ability to abate it. The standard ought to be to require the occupier to do what it is reasonable to expect him to do, given his individual circumstances.

Holbeck Hall Hotel v Scarborough BC (2000)

The plaintiff owned a hotel and grounds situated at the top of a cliff overlooking the North Sea. The defendants owned the land forming the undercliff between the hotel grounds and the sea. That part of the coast was subject to marine erosion and the cliff slopes were inherently unstable and subject to unpredictable local landslides. There had been landslides in 1982 and 1986. Some remedial work was carried out by the defendants in 1989. In 1993 a massive landslide, the size of which could not have been anticipated, caused a loss of stability to the hotel grounds and part of the hotel, which had to be demolished. The plaintiff sought damages. The Court of Appeal reversed the trial judge's decision and held that the defendants were not liable. Explaining their decision, they said that a landowner owed a measured duty of care to prevent danger to a neighbour's land due to lack of support through natural causes in circumstances in which the owner or occupier knew or was presumed to know of the hazard, even though he had not created it, and that the danger in question was foreseeable. However, a landowner would not be liable for a latent hazard simply because he would have discovered it upon further investigation. The scope of the duty of care depended not only upon the landowner's knowledge of the hazard, the ease and expense of abating it, his ability to abate it, but also the extent to which the damage that had in fact occurred was foreseen and whether it was fair, just and reasonable in the circumstances to impose the duty. In the present case justice did not require that the defendant be held liable for damage which was vastly more extensive than that which was or could have been foreseen without geological investigation and especially in circumstances in which the defect lay just as much on the plaintiff's land as on the defendant's.

Rees v Skerrett (2001)

This case involved adjoining terraced houses separated by a party wall. The plaintiff's house had the benefit of an easement of support. The defendant demolished his house and this had two consequences. Firstly, as a result of the withdrawal of support, and wind suction, some cracks developed in the now unrestrained party wall; these cracks allowed the ingress of the elements, causing damage. Secondly, and apart from the cracking just mentioned, further damage occurred to the plaintiff's property by reason of the exposure of the unprotected party wall to the elements. The plaintiff recovered damages for the damage to his house resulting from the ingress of the elements through the cracks as a result of the interference with his easement of support. However, he also recovered damages for the damage caused by the weather as a result merely of the exposure of the party wall. This was justified on the basis of an existence of a duty of care in tort, the court making reference to a number of cases including *Holbeck Hall Hotel v Scarborough BC* (above).

8
Rights to light and air

8.1 INTRODUCTION

Historically, probably as a result of written legal formulae used by conveyancers, rights of light and rights to air have been lumped together in the legal consciousness. In a physical sense they are similar in that the dominant owner benefits from something coming across another's land, as opposed to a right to go onto or positively use the other's land for some purpose – as for example with a right of way or drainage. However, on a technical level, whilst in many cases there may be some analogy of reasoning, it should not be taken too far. The two distinct rights are dealt with separately below. For a more detailed consideration of rights to light, the reader is referred to *Anstey's Rights of Light and how to deal with them* (RICS, 2006).

8.2 LIGHT

A 'right to light' is not as far-reaching as it may sound. In order to understand the limitations, it is best to start with a consideration of the precise nature of the right.

Firstly, a right to light is a right to the enjoyment of light through defined apertures *in a relevant building or structure.* There can be no right to light in respect of open land, for example, a garden or a field. Where reliance is placed upon section 3 of the *Prescription Act* 1832, only a 'dwellinghouse, workshop, or other building' will suffice; where reliance is placed upon the common law, it is generally considered that the test is whether or not the construction is a 'building'. Furthermore, not just any aperture will do: a right to light only exists in respect of light admitted through apertures whose purpose is to admit light.

Harris v De Pinna (1886)

This was a case concerning section 3 of the *Prescription Act* 1832. The plaintiffs were timber merchants, and the defendants were the occupiers of the adjoining premises. On the plaintiffs' premises were structures used for the storage and seasoning of wood, and also for exhibiting the wood to customers. These structures consisted of solid baulks of upright timber fixed in stone bases on brick piers, with cross-beams and diagonal iron braces, and were divided into floors or stagings which had open unglazed ends between the uprights and which served to admit air for drying and also light.

At first instance Chitty J held that the structure was not a 'building' within the meaning of section 3 of the *Prescription Act* 1832, and the plaintiffs' claim failed on that basis. On appeal, without deciding whether the structure was a building, the Court of Appeal said that the consequence of the nature of the structure, and the mode of carrying on business, meant that from time to time timber would be piled to block one or other of the apertures so that the plaintiffs could not prove that there had been uninterrupted access of light by any one aperture for the statutory period. Accordingly the claim failed on this basis.

Perhaps most fundamentally of all, the right only confers an entitlement to *sufficient light.*

Colls v Home & Colonial Stores Limited (1904)

The owners of a business premises on Worship Street, Shoreditch, applied for an injunction restraining the construction of a building on a piece of land across the street, believing that it would obstruct the light to their property. Whilst the proposed construction would undoubtedly have some impact on the light to the business premises, the House of Lords did not consider that this was enough to constitute an actionable interference with their right to light. Lord Lindley set out what it was that a right to light protected:

> 'generally speaking an owner of ancient lights is entitled to sufficient light according to the ordinary notions of mankind for the comfortable use and enjoyment of his

> house as a dwellinghouse, if it is a dwellinghouse, or for the beneficial use and occupation of the house if it is a warehouse, a shop or other place of business.'

It follows that, when considering whether or not there has been an actionable interference with a right to light, it is necessary to consider not how much light has been obstructed, but whether what remains is sufficient.

When considering whether sufficient light remains for the ordinary purposes of mankind, the Court will consider not just the present use of the room or building, but also any normal use to which the room or building might be put in the future.

Price v Hilditch (1930)

The plaintiff brought an action for interference with the light to his building. In particular a scullery situated in the basement was affected. The scullery had been used as such for a long time, if not since the house was built. After the obstruction of light, the amount of light remaining was sufficient for the ordinary purposes of a scullery. However, Maugham J. held that the right of the plaintiff was not limited by the use to which he had put the room, namely that of a scullery, and that the obstruction did amount to an actionable interference with the right of light to the room.

As regards internal layout, certainly in the case of a right arising by virtue of section 3 of the *Prescription Act* 1832, this is a matter of choice for the dominant owner and his rights will not be limited by reference to the present internal lay-out of the building.

Colls v Colonial (1904)

Lord Davey said that: 'The easement is for access of light to the building, and if the building retains its substantial identity, or if the ancient lights retain their substantial identity, it does not seem to me to depend on the use which is made of the chambers in it, or to be varied by any alteration which may be made in the internal structure of it.'

Carr-Saunders v Dick McNeil Associates (1986)

The plaintiff owned the second floor of a building in Neal's Yard, Covent Garden, London. It was open plan and lit by a number of windows at the front of the building, and by two windows at the rear. Later he converted the second floor into a suite of six small therapy rooms. Two of these rooms were at the rear and separated from each other by an internal partition. Each room was lit by one of the two windows at the rear. The defendants owned a property at the rear of the property in question. They constructed an additional two storeys to this property, and the plaintiff claimed that the addition of these further storeys interfered with his right to light to the rear-facing therapy rooms. The defendant argued that the effect of the construction of the therapy rooms and erection of internal partitions, which had taken place during the prescriptive period, should be disregarded, and that, when disregarded, there was no interference with the right to light. The Court disagreed with this approach, holding that the right acquired under section 3 of the *Prescription Act* 1832 related to light to the premises rather than to a particular room, and that, even before the subdivision of the second floor, it would have been necessary for the Court to consider the effect of the defendants' building works, not only in the second floor as it was then used (that is to say, as a single open space) but on any other arrangement of that space which might reasonably be expected to be adopted in the future.

What is sufficient light for the ordinary notions of mankind will depend upon the nature of the building.

Allen v Greenwood (1980)

In this case light to the plaintiff's greenhouse had been obstructed by the defendant in such a way that it could no longer be used for growing plants, although there would be sufficient light to work in it. The Court of Appeal held that what is ordinary in terms of an amount of light depends upon the nature of the building and to what use it is ordinarily adapted. Lord Justice Goff said that: 'If the building be, as it is in this case, a greenhouse, the normal use of which requires a high degree of light, then it seems to me that that degree is ordinary light.'

8.2.1 Interference with a right to light

It has already been observed above that when assessing whether or not there has been an actionable interference with a right to light, the Court will consider whether what is left after the obstruction is sufficient, rather than how much light is being obstructed. In assessing whether what remains is sufficient, the Court will take other sources of direct light (as opposed to reflected light) into account, so long as there is no reason why the dominant tenement should not continue to receive light from those sources.

Smith v Evangelisation Society Trust (1933)

The plaintiff owned a low commercial building which had the benefit of light from skylights. There came a point, during the prescriptive period, at which the roof was remodelled and this included the removal of the skylights and the construction of a window in what had previously been a gable end. A neighbour sought to build in such a way that affected the light to the building in its remodelled form. Various issues arose as to the level of light enjoyed during the prescriptive period and whether, when considering the amount of light remaining, the judge should have regard to the skylights which had been removed by the plaintiff. The Court of Appeal held that the trial judge had been correct to mentally reconstruct the premises before the remodelling and then to take account of the light through the skylights – as these skylights could not have been obstructed against the will of the plaintiff so as to render the room unfit for ordinary purposes; the plaintiff himself had chosen to do this.

Quite complicated scenarios can arise – for example, where a building or room has a double aspect, each aspect being over land of different owners.

Sheffield Masonic Hall Co. Ltd v Sheffield Corporation (1932)

In this case the Court had to consider the following conundrum: assume a building has windows on its north side and on its east side. The windows on the north side look out over Blackacre, the windows on the east side look out over Whiteacre. A right to receive light exists in respect of both the windows on the north side and the east side. There are no buildings on Whiteacre or Blackacre. Can the owner of Blackacre build as high as he likes while Whiteacre remains an open space, or is there any limit to his right? Maugham J held that once the right to receive light was acquired in respect of the windows on both aspects, a restrictive obligation was imposed upon the owners of Whiteacre and Blackacre – namely that they would not build so that by their joint action they caused an actionable interference with the light to the building. In other words, the owner of Blackacre could build to such a height as, with a similar building by the owner of Whiteacre, would still leave sufficient light for the building.

Modern methods of measuring the light to rooms are now very sophisticated. Software packages are capable of allowing surveyors to assess the impact of developments on the light to neighbouring buildings. The ability to place numbers on levels of illumination led to what is often referred to as the '50/50 rule' – the essence of which is that so long as 50 per cent of a room remains well-lit (considered to be one lumen per square foot), there is no actionable interference with the light to that room. However, this is a rule of thumb used by surveyors rather than a rule of law.

Carr-Saunders v Dick McNeil Associates (1986)

Millett J observed that 'The 50/50 rule is not in my judgment to be applied without any regard to the shape and size of the room or the disposition of the light within the room to which it is applied. The justification of the 50/50 rule is that an owner is unreasonable if he complains that the corners or other parts of the room where good light is not expected are poorly lit, if the room as a whole remains well lit.'

It is worth noting that neither planning permission nor awards under the *Party Wall etc. Act* 1996 confer immunity from private law claims for interference with an easement.

A right to light can be acquired by any of the methods described in Chapter 2. However, it is worth drawing attention to the fact that the *Prescription Act* 1832 has a separate and specific section, section 3, dealing only with rights of light:

> '3. When the access and use of light to and for any dwelling-house, workshop, or other building shall have been actually enjoyed there-with for the full period of twenty years without interruption, the right thereto shall be deemed absolute and indefeasible, any local usage or custom to the contrary notwithstanding, unless it shall appear that the same was enjoyed by some consent or agreement expressly for that purpose by deed or writing.'

It will be noted that, in contrast with section 2 of the *Prescription Act* 1832, it is not necessary for user to have been 'as of right'. Also, section 3 does not bind Crown land.

It used to be the case that in order to prevent his land becoming afflicted by a right of light to a neighbouring property, a landowner would erect a hoarding or 'spite-fence' to interrupt the light to the neighbouring property. In addition to being cumbersome, unsightly and unneighbourly, such screens would often fall foul of planning law. A solution was provided by the *Rights of Light Act* 1959, which has provided a system for the registration of notional obstructions.

8.3 AIR

When speaking of a 'right of air' one is usually speaking of a right to *receive* air to one's property over the land of another.

By contrast, no specific right as such is required to discharge air generally across another's land – although if the air is objectionable (for example, by being foul), or causes some other interference with the neighbour's enjoyment of his property, then he may have an action in nuisance. Of course, it would be a defence to any such action in nuisance to show that the discharge was authorised by reason of an easement permitting that activity; but such an easement would be more accurately referred to as a right of discharge.

Rights to receive air are usually only encountered in practice in the context of commercial buildings – typically where air is drawn in to be used in a production or drying process.

A right to receive air generally over and across adjoining undefined land cannot be acquired by prescription.

Bryant v Lefevre (1879)

The plaintiff and the defendant occupied adjoining premises. The plaintiff complained that, in rebuilding his house, the defendant had carried it up beyond its former height, checking the access of the draught of air to the plaintiff's chimneys. The Court of Appeal held that the right upon which the plaintiff relied did not exist at law and that no man could dictate to his neighbour how he should build his house with respect to the general current of air common to all mankind.

A right to receive air can however be acquired by prescription (at common law or pursuant to section 2 of the *Prescription Act* 1832) where the air is received onto the dominant tenement through a defined channel on the adjoining land, or is received through a defined aperture in the dominant tenement.

Bass v Gregory (1890)

The plaintiffs owned a public-house called the 'Jolly Anglers'. The cellar of the public-house was ventilated by a shaft cut from the cellar through rock into a disused well which stood in a yard on the land of the defendant. The cellar had been ventilated in this way for at least 40 years and with the knowledge of the occupiers of the yard. The Court was referred to *Bryant v Lefevre* (above), but held that the principle did not apply in the case of a strictly defined channel.

That said, however, under the principle known as non-derogation from grant, where a piece of land is sold for a particular purpose requiring the access of air, the party selling the land will not be permitted to use his land in a manner that prevents the land sold from being used for the purpose for which it was sold.

Aldin v Latimer, Clark Muirhead & Co (1894)

The plaintiff bought the timber business of a Mr Munro and also took a lease from him of the premises which he owned and at which he had traded. Under the lease the plaintiff was obliged to carry on the business of timber merchant at the premises. Mr Munro continued to occupy the property next door until he died, at which point it was sold to the defendants. The defendants sought to develop this property, and the plaintiff complained that in doing so they were interfering with the access of air to his drying sheds next door used in connection with his business as a timber merchant. The Court held that the lease granted to the plaintiff was specifically for the purpose of carrying on a timber business and that the passage of air to the drying sheds was necessary for this activity, and that accordingly Mr Munro had been subject to an obligation not to do anything on his adjoining property to substantially interfere with that activity. The defendants, as successors in title to Mr Munro, were bound by that obligation.

In certain circumstances, the courts may be willing to go so far as to imply an easement of necessity to allow an owner to go onto adjoining land to construct a channel to provide for the access of air to his property.

Wong v Beaumont Property Trust (1965)

The plaintiff was the tenant of a cellars operating as a popular restaurant. The lease contained covenants requiring the plaintiff to keep the premises open as a restaurant, to control and eliminate all smells and odours caused by such use of the premises, and to comply with the health regulations. In fact, the covenant to comply with health regulations could not be complied with unless a ventilation system was installed on the premises with a duct fixed to the outside back of the landlord's building. The landlord refused permission for this. The plaintiff applied to the Court for a declaration that he was entitled to enter upon the landlord's premises for the purposes of constructing, maintaining and repairing a ventilation system for use in connection with the restaurant. The Court of Appeal found for the tenant, holding that he had the benefit of an easement of necessity for the purposes mentioned. Without such an easement, the business of the restaurant could not be carried on legally and in accordance with the terms of the lease.

9 Interference with easements and remedies

9.1 INTERFERENCE WITH EASEMENTS

When does an interference with an easement entitle the person with the benefit of the easement to take action? Put another way, what amounts to an 'actionable' interference with an easement? (The remedies available to such a person are dealt with below at paragraph 9.2.)

Not every interference with an easement is an 'actionable' interference. An interference will only be actionable if it interferes substantially with the enjoyment of the easement.

Celsteel Ltd v Alton House (1986)

The lessees of flats with the benefit of rights of way across driveways and parking spaces on the ground floor level complained about the proposed construction of a car wash and a traffic flow system at that level. The Court held that an interference with their rights of way would be actionable if it was substantial, but that it would not be substantial if it did not interfere with the reasonable use of the right of way. In the circumstances, the construction of the car wash, which would reduce the driveway from a width of 9 metres to a width of 4.14 metres over a 10 to 12 metre distance, was an actionable interference, but the proposed traffic flow system was not.

West v Sharpe (2000)

In this case the Court observed that:

> 'Not every interference with an easement, such as a right of way, is actionable. There must be a substantial interference with the enjoyment of it. There is no

actionable interference with a right of way if it can be substantially practically exercised as conveniently after as before the alleged obstruction. Thus, the right of way in law in respect of every part of a defined area does not involve the proposition that the grantee can in fact object to anything done on any part of the area which would obstruct passage over that part. He can only object to such activities, including obstruction, as substantially interfere with the defined right as for the time being is reasonably required by him.'

It follows that very often the first step in any problem concerning alleged interference with an easement is to ascertain the precise scope of the easement in question. It is only then that consideration can be given as to whether the interference is in fact an actionable interference.

The next step is consideration of the impact of the interference in question upon the dominant owner's exercise of the easement. The test is one of convenience rather than the minimum requirement. For example, in the case of an impingement onto a right of way the question is one of convenience, rather than determining the minimum space necessary to exercise the right.

B&Q plc v Liverpool and Lancashire Properties Ltd (2001)

The claimants had a lease of a unit on a retail park. The defendant landlord wanted to construct a large extension at the rear of a neighbouring unit. This area behind the units was subject to an expressly granted right of way in favour of the claimants. If the extension were built, the turning movements of the claimant's vehicles would be made more difficult. Blackburne J. held that the test of an actionable interference was not whether what the grantee was left with was reasonable, but whether his insistence on being able to continue the use of the whole of what he had contracted for was reasonable, and that it was not open to the grantor to deprive the grantee of his preferred modus operandi and then argue that someone else would do things differently, unless the grantee's preference was perverse or unreasonable. He also held that the fact that an interference with an easement was infrequent and, when it occurred, was relatively fleeting, did not mean that the interference could not be actionable. It should be noted of course that this case concerned an expressly granted right of way.

9.1.1 Examples of actionable and non-actionable interferences

Placing a gate across a private right of way

This is not necessarily an actionable interference.

Pettey v Parsons (1914)

The defendant sold land adjoining his own to the plaintiff, who covenanted to construct a road on that land, along which the defendant was to have a right of way. The plaintiff constructed the road, making it ten feet wide, and placed railings along part of one side of the road, and a gate across the road, such that the defendant could not in fact use the road without going through the gate. The defendant responded immediately by removing the railing and the gate. The Court held that the private right of way had not been substantially interfered with by the erection of the railings and gate. The plaintiff was entitled to erect a gate so long as it was kept open in business hours and never locked.

Locking the outside door of a house divided into flats

This has been held not to be actionable, in a case in which each tenant had a key.

Dawes v Adela Estates Ltd (1970)

The door to a property containing flats had been fitted by the landlords with an automatic locking device as a security measure. Whenever the door closed, it locked automatically. The plaintiff claimed that this caused great inconvenience to his clients and visitors and that sometimes the post was delayed because the postman could not access the building. It was not in dispute that the plaintiff had a right of access through the entrance hall and that this right enured for the benefit of his licensees as well. The plaintiff sought an injunction restraining the defendant landlord from allowing the door to be locked between 7.00 am and 10.30 pm daily.

The Court held that it was not legitimate to derive from *Pettey*'s case (above) any principle that an outside door must always in all circumstances be left unlocked. The question was whether the locking of the door, with the provision of a key for the owner of the easement, represented a substantial interference with the easement. As regards access generally for the tenants to their flats, they had a key, and this aspect of the case did not amount to a substantial interference. The Court was concerned by the inability of the postman to obtain access and didn't feel that this difficulty could be met by the provision of a common letter box inside the front door. However, the landlord undertook to make arrangements to meet this difficulty, and on the basis of this undertaking the Court declined to grant the injunction sought.

Each case will turn upon its own facts, but note that it will not always be an answer to say that a key has been provided.

Guest Estates Ltd v Milner's Safes Ltd (1911)

The plaintiff tenant had a right of way which was blocked by the defendant landlord's locking of a gate. The defendants contended that they were entitled to lock the gates if the plaintiff was supplied with sufficient keys. The Court disagreed. In the opinion of the Court, the plaintiff had a free right of passage and it was an obstruction to keep the gate locked and it was no answer to say that keys would be supplied. It was not difficult to appreciate the inconvenience if the plaintiff were obliged to carry keys; it was impractical.

Moving a flower bed in communal gardens

Jackson v Mulvaney (2003)

The claimant lived in one of a group of cottages. Several of the occupiers of the cottages, including the claimant, used an adjoining piece of land, belonging to the defendants, as a garden. The claimant had in particular tended a flower bed on the land. The defendants, who wished to construct a gravelled path across the land, dug up the flower bed without giving any notice to the claimant. The claimant brought an action to establish an easement in respect of her use of the land and for damages. The most difficult issue in the case was how to define the extent of the easement which the claimant had acquired (by reason of section 62 of the *Law of Property Act* 1925). The Court of Appeal agreed that she had established that she had the right to use the land as a communal garden, but that right did not extend to excluding the defendants from any use which they wished to make of the land. It restricted their use only to the extent necessary to enable the land to be used as a communal garden. By removing the flower bed without notice and without giving the claimant any opportunity to recreate or relocate it or its contents elsewhere, the defendants had interfered with her easement.

Failing to clear rutted snow

This has been held not to be an actionable interference.

Clutterham v Anglian Water Authority (1986)

The failure of the owner of a road to clear compacted and rutted snow which had been rendered slippery by having partially thawed and then refrozen did not constitute an obstruction of a person's right of way.

9.1.2 Does realignment of a right of way constitute actionable interference?

Greenwich Healthcare NHS Trust v London & Quadrant Housing (1998)

The plaintiff purchased land which included a road connecting it with the public highway in order to build a hospital. Planning permission for the hospital required that a new link road and junction with the highway be built before the hospital could open. Some of the defendants had rights of way over the road. The plaintiff applied for a number of declarations, including declarations that a) the proposed realignment would not be an actionable interference because the realigned route would be equally convenient, and b) even if it was, the defendants would not be entitled to an injunction, only to damages. Lightman J granted a declaration that, in any event, the defendants would only be entitled to an award of damages. He declined to decide what he regarded as a very far-reaching question, namely whether or not the realignment would in any event be an actionable interference with the right of way.

9.1.3 Interference with light

As regards interference with light, see paragraph 8.2.1.

9.2 REMEDIES

An interference with an easement that is an 'actionable' interference will generally entitle the dominant owner to take action, perhaps through self-help, but more usually by bringing a claim before the courts for disturbance of his easement – which is a particular form of action in nuisance. These remedies are discussed below.

9.2.1 Self-help

As a matter of general principle, a person suffering a nuisance is entitled to take steps to physically remove it. This is usually referred to as 'abatement'.

R v Rosewell (1699)

> 'If H. builds a house so near mine that it stops my lights, or shoots the water upon my house, or is in any other way a nuisance to me, I may enter upon the owner's soil and pull it down.'

The extract above is from a very old case and is dangerously misleading in the context of the modern approach of the law to a) the situations in which abatement is appropriate, b) the limitations as to its exercise, and c) statutory intervention which, in certain circumstances, renders such behaviour a criminal act.

The modern approach seems to be that abatement should be confined to simple situations which would not justify the expense of legal proceedings, or where an urgent remedy is required.

Burton v Winters (1993)

The plaintiff had established at trial that the garage built by the defendants' predecessors in title encroached onto her property by about 4½ inches, but the judge had declined to grant her a mandatory injunction requiring removal of the encroaching part of the garage. The plaintiff appealed the refusal of the injunction and lost in the Court of Appeal. Her petition to the House of Lords was refused. She was unwilling to accept that this was the end of the matter and embarked on a campaign of trespass, including building a wall in front of the defendants' garage! As a result of her conduct the defendants sought and obtained an injunction restraining her from trespassing on their land. She breached this and was ultimately committed to prison for two years. In the course of considering the appropriateness of the sentence of imprisonment, the Court of Appeal were concerned that at some point the plaintiff might seek to demolish the encroaching part of the garage relying upon the common law right of abatement. The Court commented that for a long time the courts have confined the remedy by way of self-redress to simple cases such as an overhanging branch, or an encroaching root, which would not justify the expense of legal proceedings, and urgent cases which require an immediate remedy. Even if the plaintiff had acted promptly, this would not be an appropriate case for abatement.

Where exercise of the right to abate is appropriate, great care must be taken in exercising it. For example, the party exercising the right must go no further in his actions than is necessary to abate the nuisance, and if he does so he exposes himself to an action for trespass and possibly prosecution (see below).

Lagan Navigation Co. v Lambeg Bleaching, Dyeing & Finishing Co. (1927)

After a heavy flood, the respondents cut away the bank of a canal to allow the flood water to escape. They sought to justify their actions on the basis that they were abating a nuisance, namely the earlier raising of the banks by the appellant. The Court considered that the raising of the banks did not constitute a nuisance, but even if it had done so, the course of conduct pursued by the respondents was not justified.

Generally, other than in an emergency, notice must be given before going onto the servient tenement to abate a nuisance. Even where it is not necessary to go onto the servient tenement, it has been suggested that it would be desirable to give notice.

Lemmon v Webb (1894)

The defendant lopped off boughs of trees to the extent that they overhung his land. The trees stood on the plaintiff's land. The plaintiff brought an action claiming that the defendant had no right to lop the trees, as they had overhung the defendant's land for more than 20 years. At first instance the Court found that the defendant was entitled to lop the trees, but not without first giving notice to the plaintiff. On appeal the Court held that as the defendant had not gone onto the plaintiff's land he had not needed to give notice to the plaintiff before lopping the trees. However, it deprived the defendant of his costs, saying that 'No-one but an ill-disposed person would do such an act without previous notice. There was no emergency in this case. The defendant has acted in an unneighbourly manner, and I cannot help thinking he intended to cause annoyance.'

Perhaps most seriously of all, a person who purports to exercise a right to abatement may find himself charged with criminal damage pursuant to section 1(1) of the *Criminal Damage Act* 1971 – although it is a defence if the party can establish that he had a 'lawful excuse'. Moreover, where there is entry into a building on the servient tenement, the would-be abator may find himself at risk of a charge of burglary (one of the circumstances in which the offence of burglary is committed is when there is unlawful entry into another's building with the intention of causing criminal damage).

9.2.2 Court proceedings

In almost all cases of any significance, the appropriate step will be to seek a remedy through the courts by bringing an action for nuisance.

Who can sue?

An action for nuisance can only be brought be a person with a right in the land affected which, in the context of easements, means a person with a right in the dominant tenement.

Hunter & Others v Canary Wharf Ltd & Others (1995)

The plaintiffs had brought an action in nuisance and negligence against the defendants arising out of construction works at a well-known development in London. The claim in nuisance concerned an alleged interference with television reception, and also in respect of dust created by the construction of a link road. A great number of the plaintiffs were not persons who had a right to exclusive possession of the properties affected, but were variously the husbands, wives, partners or children or other relatives of such persons. The House of Lords reviewed the case law on nuisance and concluded that the right to sue in nuisance did not extend to so wide a category of persons and that, generally, only a person with an interest in the land affected could bring a claim in nuisance.

It is also necessary that the person bringing the action has the benefit of the relevant easement accommodating the dominant tenement. An example of a situation in which a party might have a right in the dominant tenement without having the right use the easement in question is the tenant of a property who, by reason of the terms of his lease, has not been granted the right to use a particular right of way serving the property.

Furthermore, a tenant can only bring an action for nuisance if the tenancy is vested in him at the time of the nuisance. A landlord can only bring a claim for nuisance where there is or may be damage to his reversionary interest – so, for example, a temporary interference with an easement presently benefiting only his tenant for the time being may not be actionable by a landlord.

Baxter v Taylor (1832)

The plaintiff owned the freehold of a close, but had let it out to a tenant. The defendant had entered upon the close with horses and cart, and, although asked to cease from doing so by the plaintiff, had continued to do so, claiming a right of way. To the extent that there was any damages to the land it was the impression of the horses and cart made on the soil. The Court of Appeal considered that this was damage of a transient nature only, rather than damage to the plaintiff's reversionary interest, and there being no injury to the plaintiff's reversionary interest, he could not bring the claim in nuisance.

Who can be sued?

The party actually interfering with the easement, whether or not he has any interest in the servient tenement, can be sued

Thompson v Gibson (1841)

A building had been built, under the superintendence of the defendants, on land which belonged to the corporation of Kendal but upon which a market was lawfully entitled to be held, the plaintiff being the owner of the market. The plaintiff brought an action in nuisance against the defendants, claiming an interference with his rights. On behalf of the defendants it was contended that they were not responsible for the continuance of the nuisance as they were distinct persons from the corporation, and whilst guilty of erecting the building, they could not be considered to have continued the nuisance as they were not in possession of the land upon which the building was built, and moreover that it would be wrong to make them liable as they could do nothing now to remove the building. The Court held that they could be liable for the continuing nuisance.

In addition to the party causing the nuisance, anyone who authorises or continues the nuisance may be liable for the nuisance.

Injunctions

An injunction is an order of the Court directing a person to do or to refrain from doing something. Breach of an injunction is punishable as a contempt of court and may result in a fine or a prison sentence or sequestration of assets. The grant of an injunction is an exercise of discretion by the Court, although there are well-established principles which guide the exercise of that discretion. Relevant matters include whether or not the nuisance is serious as opposed to trivial, whether it can in fact be compensated in damages, and whether the applicant has waited a long time before bringing his claim or has somehow acquiesced in the nuisance. Depending upon the circumstances, the Court may not be prepared to grant an injunction, or it may choose to award damages instead of an injunction.

Interim injunctions

Injunctions can be granted on a temporary ('interim') basis until such time as a court can make a final determination as to the rights of the parties (for more detail see *Civil Procedure Rules* Part 25, and the Practice Direction on Interim Injunctions). They can be granted very quickly – by telephone if necessary, and even (in the right circumstances) before the other party is notified that an application for an injunction has been made.

An important aspect of an application for an interim injunction is the 'undertaking in damages'. If the Court is prepared to grant an interim injunction, the price that the applicant must usually pay is to give an 'undertaking in damages'. This is a promise to the Court to pay compensation to the other party if, when the matter has been considered at a final hearing, the Court decides that the applicant was not entitled to an injunction after all and considers that he should pay compensation to the respondent.

It should be appreciated immediately that the compensation payable might be very significant, for example where building works have been halted by an interim injunction that was unwarranted. In many cases it will be reasonably clear that the claimant is entitled to an injunction, and so the fact that he has to make this promise in order to obtain an injunction at the interim stage will not worry him unduly. In other cases, where it is less clear, the undertaking may be of a greater concern.

In deciding whether or not to grant an injunction at the final hearing, one of the matters to which the Court will have regard is whether or not the claimant has acted promptly in seeking an injunction, and in particular whether or not he applied for, or should have applied for, an interim injunction.

Gafford v Graham (1999)

The defendants were required by reason of a covenant to submit any building plans for the approval in writing of their neighbour, the claimant, before commencing to build. Between 1984 and 1988 the defendant built in breach of this covenant. By the time of the trial the building had stood for over seven years. The plaintiff had not applied for an interim injunction. The Court of Appeal upheld the county court judge's refusal of an injunction, stating that 'as a general rule, someone who stands by knowing that he has clearly enforceable rights while a permanent and substantial structure is unlawfully erected ought not to be granted an injunction to have it pulled down'.

However, it would be unwise to assume that a failure to apply for an interim injunction will always be fatal.

Mortimer v Bailey (2004)

The defendants started to build an extension to their house in breach of a restrictive covenant in favour of the claimants. The claimants did not apply for an interim injunction until almost two months after the commencement of the building works, by which time there was only about seven days' work left to be done. The county court judge refused an interim injunction, but at trial some time later granted the claimants a final injunction directing the defendants to demolish the extension; alternatively to make certain alterations to it. The defendants appealed, contending that the judge should not have granted a final injunction because of the claimants' delay in applying for the interim injunction. They relied upon a statement of the court in *Gafford v Graham* referred to above. The Court emphasised that each case turned on its facts and, moreover, the grant or refusal of an injunction was always an exercise of discretion. In the context of the quotation from the decision in *Gafford v Graham* above, Gibson LJ observed that he had some doubt as to whether it was appropriate to say that a person who does not proceed for an interim injunction when he knows that a building is being erected in breach of covenant, but who has made clear his intention to object to the breach and to bring proceedings for that breach, should generally be debarred from obtaining a final injunction to pull down the building. He thought that there may be many circumstances in which a claimant would not be able to take the risk of seeking an interim injunction, but agreed that not to seek an interim injunction is a factor which can be taken into account in weighing the balance whether a final injunction should be granted.

Damages in lieu of an injunction

The grant of an injunction is a discretionary remedy. In certain cases the Court will decline to grant an injunction but will award damages in place of an injunction. When will it do this?

Shelfer v City of London Electric Light Co. (1895)

In dismissing the plaintiff's claim for an injunction and granting damages in lieu, the Court of Appeal set out a 'working rule' as to when damages might be awarded in substitution for an injunction, namely cases in which the following four requirements existed:

- where the injury to the innocent party was small; and
- the injury could be estimated in terms of money; and
- the innocent party could be properly compensated by a small payment of money; and
- it would be oppressive to grant an injunction.

The result of these guidelines was that it became the exception rather than the rule to award damages in place of an injunction. However, the case of *Shelfer v City of London Electric Light Co.* was decided in the nineteenth century. In the last 16 years or so – principally since another well-known case, *Jaggard v Sawyer* – there seems to have been a change in emphasis, and damages have been awarded in place of injunctions in situations in which, previously, one would have expected an injunction to have been granted. It is difficult to shepherd together the examples of such situations and to extract clear guidelines, but it does seem that more and more, the courts will be concerned with whether or not an injunction would be oppressive. In this context, the delay (or otherwise) of the innocent party in seeking relief continues to be an important factor.

Jaggard v Sawyer (1995)

In this case the Court of Appeal upheld the county court judge who had declined to grant the plaintiff a final injunction. The Court of Appeal endorsed the working rule in *Shelfer* (above), and considered that the judge had been right to apply it in the case, but commented that: 'it is only a working rule and does not purport to be an exhaustive statement of the circumstances in which damages may be awarded instead of an injunction'.

Midtown Ltd v City of London Real Property Co. Ltd (2005)

In this case the Court was faced with a claim for an injunction, alternatively damages in lieu of an injunction, in connection with a right to light. Reviewing the authorities, Peter Smith J said that the judgments establish a willingness on the part of the courts to depart from the strict requirements set out in *Shelfer* in an appropriate case. He declined to grant an injunction to the freeholder of the affected building, giving four reasons: i) it was only interested in the property from a money-making point of view, i.e. as an investment, ii) there was probably no loss to the freehold, iii) it probably had development plans of its own which would make any injunction academic, and iv) in the Court's view the defendants had behaved reasonably, flagging the issue in correspondence and seeking meetings, but were rebuffed. He also commented that it would be oppressive to prevent the defendants from pursuing a worthwhile and beneficial development for the area. The leaseholder's claim for an injunction was also refused, the judge commenting that they had not responded to the defendant's open approach, the infringement of their rights would be even less (because they had no real capital interest in the building and there would be no effect on their use of the property), and the reality was that modern offices used, as this one did, artifical light to maintain a constant light.

Regan v Paul Properties Ltd (2006)

This was another case involving interference with a right to light. At first instance the Court declined to grant an injunction. Giving judgment on appeal, Mummery LJ referred to a number of propositions which he derived from the judgments in *Shelfer*, none of which, he considered, had been overruled by later decisions:

- A claimant is prima facie entitled to an injunction against a person committing a wrongful act, such as continuing nuisance, which invades the claimant's legal right.
- The wrongdoer is not entitled to ask the court to sanction his wrongdoing by purchasing the claimant's rights on payment of damages assessed by the court.
- The court has jurisdiction to award damages instead of an injunction, even in cases of a continuing nuisance; but the jurisdiction does not mean that the court is 'a tribunal for legalising wrongful acts' by a defendant, who is able and willing to pay damages.
- The judicial discretion to award damages in lieu should pay attention to well-settled principles and should not be exercised to deprive a claimant of his prima facie right 'except under very exceptional circumstances.'

Although it is not possible to specify all the circumstances relevant to the exercise of the discretion or to lay down rules for its exercise, the judgments indicated that it was relevant to consider the following factors: whether the injury to the claimant's legal rights was small; whether the injury could be estimated in money; whether it could be adequately compensated by a small money payment; whether it would be oppressive to the defendant to grant an injunction; whether the claimant had shown that he only wanted money; whether the conduct of the claimant rendered it unjust to give him more than pecuniary relief; and whether there were any other circumstances which justified the refusal of an injunction.

Site Developments (Ferndown) Ltd v Barratt Homes Ltd (2007)

The claimants asserted ownership of a 'ransom strip' controlling access to an area of land which had been developed and upon which there now stood 34 houses. The claimants had not threatened or sought an interim injunction, but had asked for proposals and negotiations. The defendants made an application for summary judgment to determine that the claimants would not, in any event, be entitled to an injunction. The Court considered that, while the position of the claimants should be considered as indicated by the first three criteria mentioned in *Shelfer* (above), the key question is whether the grant of an injunction would be oppressive to the defendant, and whether it would be oppressive depends on all the relevant circumstances existing as at the date the Court is asked to grant an injunction. In this case it would be oppressive and summary judgment was given on this part of the claim in favour of the defendant.

Measure of damages

In many situations a party will, in addition to seeking an injunction to restrain future nuisance, also seek damages for past nuisance. In such a case the Court will do its best to put a figure on the loss that has been sustained by the injured party.

In cases where the Court exercises its discretion to award damages in lieu of an injunction, the measure of damages awarded may reflect the amount which the parties would (viewed objectively) have agreed as the price of the right to continue the nuisance. This is sometimes referred to as 'negotiating damages', and reflects the fact that the dominant owner is being compensated not so much for any loss occasioned by the actual physical interference with his easement but the loss of bargaining power to agree financial compensation (by reason of the refusal of an injunction).

Wrotham Park Estate Co v Parkside Homes Ltd (1974)

The Court declined to grant a injunction requiring demolition of houses built in breach of a restrictive covenant. The Court held that damages should be such a sum as the plaintiffs might reasonably have demanded as a quid pro quo for relaxing the covenant, had the defendants asked them. In all the circumstances it would be right to award them 5 per cent of the reasonably anticipated profits of the developer.

Lunn Poly Ltd v Liverpool & Lancashire Properties Ltd (2006)

At trial, the tenant of a unit in a shopping centre was granted damages in lieu of an injunction restraining the landlord from breaching its covenants by relocating a fire door without the tenant's consent. A question arose as to the basis of assessment. The Court of Appeal held that the courts are not limited to any specific basis for assessing damages in lieu of an injunction under the Act, but, principle and practice suggested that the normal three bases would be: (a) traditional compensatory damages, i.e. a sum which compensates the claimant for past present and future losses as a result of the breach but not for the loss of the covenant; (b) negotiating damages, i.e. a sum based on what reasonable people in the position of the parties would negotiate for a release of the right which has been, is being, and will be breached; and (c) an account, i.e. a sum based on an account, that is, on the profit the defendant has made, is making and will make as a result of the breach. In relation to these three types of assessment, one would generally expect the normal approach adopted by the courts to be applied. Thus, one would normally expect that damages under (a) and (b) would be assessed at the date of breach, that any such damages would not be punitive, and that damages under (b) would be assessed by reference to facts as they were at the valuation date. However, there are no absolute rules.

10 Miscellaneous matters relating to easements

10.1 EASEMENTS AND ANCILLARY RIGHTS

Generally speaking, the grant of an easement carries with it the grant of an entitlement to do what is reasonably necessary to enjoy the easement.

Bulstrode v Lambert (1953)

The plaintiff, who carried on business as an auctioneer, had a right of way across the defendant's yard and to his auction mart. The precise terms of the conveyance which created the right of way were: 'the right to pass and repass with or without vehicles over and along the land coloured brown on the said plan for the purpose of obtaining access to the building at the rear of the said premises and known as the auction mart'. Whilst it was possible to drive large vehicles onto and across the defendant's yard, thereafter the accessway to the auction mart was too narrow, and furniture and other goods had to be taken the remainder of the way on foot. The plaintiff claimed an entitlement, ancillary to the right of way, to stop his vehicles on the yard for so often and so long as might be necessary for loading and unloading persons or goods proceeding to or from the auction mart.

The Court held that it was a matter of construing the grant, and that in this case the whole object of the grant was for the purpose of the grantor obtaining access to the auction mart. The Court held that the plaintiff was entitled under the grant to bring goods to the auction house by vehicle for auction and sale, and that if he could do that, he must of necessity be entitled to unload them.

10.1.1 There are limits

VT Engineering Ltd v Richard Barland & Co. Ltd (1968)

The plaintiffs had the benefit of a right of way over a road on the defendants' property. They used the road for moving heavy engineering goods to and from their premises. The defendants proposed to construct a building in such a way as would not allow the plaintiffs any lateral 'swing room' when transporting goods along the road. The plaintiffs brought an application for an injunction to prevent this alleged interference with the right of way. The Court considered that a grantor of a right of way in such a case might have to allow some degree of tolerance for wide loads, particularly if there were bends, but that was different from what the plaintiffs sought, the effect of which would be to sterilise a strip of land of indefinite depth on each side of the way.

10.1.2 The extent of the ancillary right will depend upon the circumstances

Nationwide Building Society v James Beauchamp (2001)

This case concerned a development of ten plots to be accessed by a common estate road. The grantor had granted a 'right of way at all times and for all purposes over the estate road constructed, or to be constructed, on the road laid until taken over by the local authority'. It was intended that once made up, the estate road would be adopted by the local authority and it was a part of the agreement between the relevant parties that the developing company would in fact make the estate road up to an adoptable standard. The developing company did not complete the estate road, and the plaintiffs, who had the benefit of the right of way, argued that in the circumstances they were entitled, by way of ancillary right, to complete the road to an adoptable standard. The Court observed that the grant of a right of way is to be taken to carry with it such ancillary and incidental rights as are necessary to make the grant fully effective, and that this is so whether the right of way is obtained by prescription or by express grant. It held that, in this case, the correct starting point was to consider the question of what right of way was granted against the background of the transaction as a whole as the extent of the ancillary right must be determined in the light of the particular circumstances of the right of way. In the present case, the parties themselves had specified the standard to which the road should be constructed, and this was determinative.

In the case of a right arising by prescription, the extent of the ancillary right will necessarily depend upon the nature and extent of the usage giving rise to the easement by prescription

Mills v Silver & Others (1991)

The first and second defendants owned a hill farm which, when it was dry enough, could be accessed by vehicles via a track over the plaintiff's land. They engaged the third defendant to lay a stone road on the track to make it passable in all weathers. The plaintiff sought an injunction to prevent the work. On appeal, the defendants succeeded in establishing a right of way along the track to their farm arising by prescription, but the Court dismissed their claim to be entitled to pave the track. It reasoned that as the owner of land with the benefit of a prescriptive right along a track is not entitled to increase the burden on the burdened land by building further buildings, there was no reason why he should be entitled to do so by making a road over the burdened land so as to make it usable at all times of the year and in weather conditions when it was not passable before. A prescriptive right of way differed from a right of way by express grant, in that the extent of a prescriptive right of way was limited by the nature of the user from which it arose.

10.2 EXPRESS GRANTS

10.2.1 Grants qualified by reservation and development clauses

Sometimes a right is granted subject to a qualification. The most common is perhaps the grant to a tenant in a block of flats of the right to use the common parts subject to the rights of other tenants to do the same – a necessary practical step for the enjoyment of all. In many instances, however, the grantor qualifies a right granted, directly or indirectly, by reference to a right he is reserving to *himself*. This often arises, and is particularly important, in the context of development land. So, for example, a person may sell part of their land to a developer, but wish to retain for themselves and their successors an unfettered ability to build upon their own retained land. Unless they can do so effectively, then in due course their land may lose value (as, for example, when after 20 years the developed land next door has acquired rights of light and air). Vendors commonly seek to achieve this objective by including a 'development clause' within the conveyance or transfer.

Where rights have been granted, but qualified by a reservation of rights to the grantor, the approach of the courts has been, where possible, to construe the reservation as permitting what would otherwise amount to an actionable interference with the rights granted, so long as the effect is not to entirely destroy the benefit of the rights granted.

Overcom Properties v Stockleigh Hall Residents Management Ltd (1989)

The Court had to construe leases which granted the tenants rights of access over the grounds and forecourts of a block of flats, but reserved to the landlord the right to develop, 'notwithstanding that the access of light or air or any other easement appertaining to the flat may be obstructed or interfered with'. The landlord wanted to mark out parking spaces and insert lockable posts to regulate the access to the spaces; this would cause a substantial interference with the rights of access granted to the tenants. The Court held that looking at the lease as a whole, the situation of the flat and the entrances, the words 'obstructed' or 'interfered with' should be read as permitting acts which would otherwise be an unjustified interference with an easement, but not acts which would for practical purposes destroy it. In the circumstances the landlord was entitled to carry out the proposed work.

In other situations the purpose of a development clause may be to prevent the acquisition of rights, rather than to restrict rights expressly granted.

RHJ Ltd v F. T. Patten (Holdings) Ltd (2007)

A clause in a lease of a property expressly excluded the implied grant of any easement of light. At the time of the action the property had enjoyed light for 20 years and the claimant contended that a right to light had arisen by reason of section 3 of the *Prescription Act* 1832. The Court had to consider whether this could be so, given a development clause in the lease which reserved to the landlord the 'full and free right to erect build rebuild and/or alter' property on his adjoining land, and given that the *Prescription Act* 1832 provides that no right to light arises where it appears that the light has been 'enjoyed by some consent or agreement expressly made or given for that purpose by deed or writing'. Lewison J considered that a development clause would satisfy the requirements of section 3 of the *Prescription Act* 1832 and prevent the acquisition of a right to light if what it authorised would interfere with light. It was not necessary to expressly refer to 'light' in the development clause, nor to provide that the enjoyment of the light was permissive.

10.2.2 The rule against perpetuities

Another matter which is sometimes encountered in conveyances and transfers, especially in the context of development land, is the grant *prospectively* of easements – often rights of way, or drainage, or for the running of services. In other words, the conveyance or transfer may grant rights over paths, roads, drains or service ducts which have not, at the time of the grant, actually been built. Depending upon the words of the grant and the relevant circumstances, such a grant might be a grant of a future rather than a present interest.

At common law, in many cases, such future interests were void under the so-called 'rule against perpetuities' – a rule developed to reflect the law's policy of not permitting property rights to be dictated many years in advance.

Dunn v Blackdown Properties Ltd (1961)

Two plots of land had been conveyed by conveyances dated 17 December 1926 and 11 January 1938, together with the express grant of a right to use the sewers and drains 'now passing or hereafter to pass' under a private road. In fact, at the time of the grant, there were no sewers or drains under the road, although at some time after the grant a surface water sewer was constructed and ran beneath the road. The plaintiff who was a successor in title to the plots claimed a right to use the surface water sewer under the private road, now owned by the defendants. The Court held that, as no sewer was in existence at the date of the conveyance to the plaintiff's predecessors in title, the grant of the right to use the sewers and drains 'hereafter to pass' was the grant of an easement to arise at an uncertain date in the future not limited to take effect within the perpetuity period, and was therefore void.

The position at common law has been considerably relaxed by statute.

Section 162(1) of the *Law of Property Act* 1925 declares that: 'the rule of law relating to perpetuities does not apply and shall be deemed never to have applied to any grant, exception or reservation of any right of entry on, or user of, the surface of land or of any easements, rights, or privileges over or under land' for the purpose of a wide variety of activities relevant in the context of easements and development land, including for example, 'executing repairs alteration or additions to adjoining land, or the buildings and erections thereon' (s. 162(1)(d)(iii), LPA 1925), and the 'constructing, laying down, altering, repairing, renewing, cleansing and maintaining of sewers, watercourses, cesspools, gutters, drains, water-pipes, gas-pipes, electric wires or cables or other like works' (s. 162(1)(d)(iv), LPA 1925).

However, some care is required when applying the section: it has been held for example that the section is directed at ancillary rights necessary to give effect to substantive easements.

Dunn v Blackdown Properties Ltd (1961)

In this case, referred to above, it was also argued that the right to use sewers 'hereafter to pass' was saved by section 162(1)(d)(iv) of the *Law of Property Act* 1925. The Court disagreed, and held that whilst wording of this section was apt to save an otherwise void ancillary right (e.g. a right to enter to carry out maintenance), it did not assist in circumstances where the substantive right itself (i.e. the right to use the sewer) was void.

For grants after 15 July 1964, section 3 of the *Perpetuities and Accumulations Act* 1964 effectively disapplies the rule against perpetuities for a period known as 'the perpetuity period' – and so as long as the interest crystallises within the perpetuity period, it will be saved. The perpetuity period in any given case will depend upon rules of common law, and/or section 1 of the 1964 Act and the words of the instrument creating the easement.

10.3 EASEMENTS AND ADVERSE POSSESSION UNDER THE LAND REGISTRATION ACT 2002

Where a squatter obtains title to a property by adverse possession pursuant to the *Land Registration Act* 2002, he will be bound by easements already affecting the land.

LRA 2002, Schedule 6 paragraph 9(2): '[Subject to...] the registration of a person under this Schedule as the proprietor of an estate in land does not affect the priority of any interest affecting the estate.'

It seems likely that a squatter who obtains title to a property by adverse possession pursuant to the LRA 2002 will also enjoy the benefit of any easements existing for the benefit of the freehold, as LRA 2002, s. 11(3) provides that the effect of registration has amongst other things the effect of vesting in the registered owner 'all interests subsisting for the benefit of the estate'.

10.4 PROTECTION OF EASEMENTS

Since the *Land Registration Act* 1925, Her Majesty's Land Registry has kept a register of title to land in England and Wales. The vast majority of land in England and Wales is now held under titles appearing on this register, and such land is known as 'registered land'. In most cases it is necessary to take formal steps to protect existing easements if they are to be preserved for the benefit of successors in title, and this is done, in the case of registered land, by registering the easement on the titles affected.

This area of property law has recently undergone some change with the introduction of the *Land Registration Act* 2002, which replaced the *Land Registration Act* 1925. It is a complicated area, and what follows here is only an outline. It should be appreciated that, should it be necessary to trace the history of an easement through the hands of owners down through the years, it is likely to be necessary to consider the law as it was under the *Land Registration Act* 1925, and sometimes earlier; it may also be necessary to consider the rules relating to unregistered land. For a more detailed treatment of these subjects the reader is referred to Ruoff and Roper's *Law and Practice of Registered Conveyancing*.

10.4.1 Easements and the Land Registration Act 2002

A legal easement which is protected by registration will exist indefinitely, binding both the land which enjoys the benefit of the easement and the land affected by it: for example, if A's land enjoys a right of way over B's land, and this right of way has been registered, if A sells his land the new owner will have the benefit of the right of way. Conversely, if B sells his land, the new owner will be bound by the right of way.

Where an unregistered freehold or leasehold estate is registered for the first time, there is a specific set of rules dealing with what unregistered interests will bind the first registered proprietor: see Schedule 1, *Land Registration Act* 2002. Interests which will override first registration and are relevant to this book include a legal easement or profit, customary rights, and public rights such as public rights of way.

Where a legal easement is not registered, the position is more complicated. An unregistered legal easement created by express grant or reservation *after* the coming into force of the relevant provisions of the LRA 2002 will generally not override the sale and registration of the land burdened by the easement. However, an unregistered legal easement in existence when LRA 2002 came into force will generally have overridden such transactions (under the transitional provisions, see Schedule 12 paragraph 10) taking place before 13 October 2006. After that they will only survive such transactions in limited circumstances.

The position of equitable easements is even more precarious. They will generally only survive the sale and registration of the burdened land if they have been protected by registration. They will not override first registration.

A proprietor claiming the benefit of an easement by prescription may make an application to the registrar for the easement to be noted on the register. The proprietors of any affected registered land will be notified and may object. Sometimes the matter becomes litigious. Guidance on making an application to the registrar may be found in the Land Registry Practice Guide (LRPG052) *Easements claimed by prescription,* available to download without charge online at www.landregistry.gov.uk

11
Statutory rights

11.1 THE PARTY WALL ETC. ACT 1996

11.1.1 Introduction

The *Party Wall etc. Act* 1996 came into force across England and Wales on 1 July 1997 (subject to some transitional provisions), and extended to almost all of England and Wales a regime that had applied in one form or another to most of London and Bristol since the nineteenth century. It is a regime for regulating the affairs of neighbours at and near the junction of their properties – note the 'etc.' in the title – its scope is wider than merely party walls.

In addition to regulation, the Act also confers rights which would not necessarily otherwise exist at common law – and for that reason it deserves a brief mention in this book (for a more detailed consideration of the Act, the reader is referred to *Case In Point: Party Walls* by Sarah Hannaford and Jessica Stephens.

11.1.2 Work which engages the Act

Broadly speaking, there are three categories of work which engage the Act. It is not possible to describe each category accurately and exhaustively in a few words, and so the short descriptions which follow are 'working' descriptions – useful shorthand, if not actually 100 per cent accurate:

- section 1 works – building work at the line of junction of adjoining properties;
- section 2 works – works to a party (fence) wall;
- section 6 works – excavation (and construction) within a specified distance of a neighbour's structure.

Various terms, including party fence wall, party wall and party structure are defined in section 20 of the Act.

Note that virtually all of the principal sections in the Act have been copied, with little or no amendment, from the *London Building Acts*. Accordingly, case law on those Acts remains relevant to questions of interpretation of the language and section of the Act, and also the overall scheme of the Act – a number of such cases appear below.

11.1.3 Notices

Except in certain limited circumstances, the Act requires proper service of the relevant notices before works in the above categories (see 11.1.2 above) are permitted to proceed.

London and Manchester Assurance Co. Ltd v O & H Construction (1989)

The parties owned adjacent plots of land separated by a party fence wall. The respondents had behaved badly, demolishing the party fence wall and had committed acts of trespass by building over the boundary line and allowing the boom of their crane to oversail the applicant's property. The work had required service of notice pursuant to the *London Building Acts (Amendment) Act* 1939 (see 11.1.2 above), but no notice had been served. Of the failure to serve notice, Harman J said: 'So far as concerns the building in breach of the *London Building Acts*...it is just as bad a case of invasion of a legal right, to tear down a party wall and put up your own wall in flagrant defiance of the most important provisions of the *London Building Acts*...as to commit a straightforward trespass'. On this basis and on the basis of the trespasses, a mandatory interim injunction was granted requiring demolition of the offending work.

Depending upon the nature of the proposed works, the notices must be accompanied by certain documents/details. The approach of the courts has generally been to require strict compliance with the terms of the Act – and this seems to apply equally in respect of any requirements relating to notices, including requirements relating to the content of the notices and the person(s) serving and the person(s) receiving notices.

Hobbs, Hart & Co. v Grover (1899)

The applicant and respondent owned adjoining properties separated by a party wall. The respondent wanted to demolish and rebuild his property and served a notice under the section in the *London Building Act* 1894 (see 11.1.2 above) equivalent to section 3 of the 1996 Act. The notice should have specified the 'nature and particulars of the proposed work' (this requirement appears in the 1996 Act at section 3(1)(b)). Instead it simply said that the intention was 'to execute such of the following works to the said party structure as may on survey be found necessary or desirable', and listed all of the permitted works set out in the section of the *London Building Act* 1894 (the equivalent section of section 2 of the 1996 Act). The applicant applied for an injunction to restrain the works and the judge ordered the respondent to provide further information about the intended works. The respondent appealed, but after unfavourable indications from the Court of Appeal settled the appeal. The Court of Appeal had observed, amongst other things, that 'the notice ought to be so clear and intelligible that the adjoining owner may be able to see what counter-notice he should give to the building owner under section 89 [of the *London Building Act* 1894]. This is the key to the whole matter.'

Spiers & Son Ltd v Troup (1915)

In this case the plaintiff was a contractor who intended to enter into a building agreement with the freeholder of a site, and then to develop it. The defendant owned the property next door. The development works required service of a notice under the *London Building Act* 1894 (see 11.1.2 above). The plaintiff served a notice which was vague in a similar way to the notice served in *Hobbs, Hart & Co. v Grover* (above), and accordingly invalid. However, there was another reason why the notice was, in the Court's opinion, invalid. At the time the notice was served, the plaintiff was not an 'owner' within the definition of the Act, having not signed the building agreement.

11.1.4 Disputes

Not infrequently, property owners refuse to consent to the proposed works contained in a notice served under the Act and a 'dispute' arises. Indeed, if an owner fails to respond at all to a notice, they are deemed to have refused to consent and there is a deemed 'dispute'.

The Act provides a statutory mechanism for the resolution of disputes by a tribunal of appointed/selected 'surveyor(s)' (see generally section 10) – although there is no requirement that these people actually be surveyors, and often they are architects or engineers; in fact, there is no requirement that they have any professional qualifications at all.

The surveyors resolve disputes by publishing an 'award'. An award can be appealed to the county court, but there are strict time limits (section 10(17)):

> '10(17) Either of the parties to the dispute may, within the period of fourteen days beginning with the day on which an award made under this section is served on him, appeal to the county court against the award and the county court may:
>
> (a) rescind the award or modify it in such a manner as the court thinks fit; and
>
> (b) make such order as to costs as the court thinks fit.'

11.1.5 Jurisdiction of the surveyors

The jurisdiction of the surveyors to resolve disputes comes from the Act. Whilst the jurisdiction of the surveyors is very wide (see section 10(10) and 10(12)), if they stray outside of their jurisdiction, the award, or the relevant part of the award, will be of no effect.

Gyle-Thompson v Wall Street (Properties) Ltd (1974)

This case concerned the *London Building Acts (Amendment) Act* 1939. Unlike the 1996 Act, there was no right under the 1939 Act for a building owner to reduce the height of a party wall. An award was made authorising the reduction in height of the party wall. The time for appealing under the Act expired and the defendants asserted that they were entitled to carry out the works. The Court disagreed. Brightman J said:

> 'In my judgment this submission is not correct in relation to an award which is ultra vires and therefore not a valid award. In the present case the defendants claimed a right which, in my judgment, they did not have, namely, a right to reduce the height of a party fence wall, and the two surveyors made an award which, in my judgment, they had no power to make. In my view the plaintiffs are entitled, in those circumstances, to come to this court to prevent a wrongful interference with their property.'

11.1.6 Interface with common law rights

The scheme of the Act is, in the case of certain works, to override common law rights

Louis v Sadiq (1997)

The defendant commenced works to the party wall between his property and the plaintiff's without service of a notice under the *London Building Acts (Amendment) Act* 1939 (see 11.1.2 above). The works caused damage to the plaintiff's property. The plaintiff applied for and obtained an injunction halting the works on the basis that they amounted in law to a nuisance. At a later hearing on damages the Court of Appeal made observations about the interaction of the Act and the common law. Referring to earlier authorities, Evans LJ said:

> 'These authorities establish, in my judgment, that the appellant would not have been liable in nuisance if he had given notice, or obtained consent, in accordance with the Act and then done no more than was agreed or was approved by the surveyors.... The adjoining owner's common law rights are supplanted when the statute is invoked, which can have the effect of safeguarding the building owner from common law liabilities when he complies with the statutory procedures, just as he may incur liabilities under the statute which did not exist at common law.'

However, note that:

- in connection with section 6 works (excavation and construction within a specified distance of a neighbour's structure), subsection 6(10) provides that:

  '6(10) Nothing in this section shall relieve the building owner from any liability to which he would otherwise be subject for injury to any adjoining owner or any adjoining occupier by reason of the work executed by him.'

  and
- in relation to all works, section 9 provides that:

  '9. Easements

  Nothing in this Act shall–

  (a) authorise any interference with an easement of light or other easements in or relating to a party wall; or

  (b) prejudicially affect any right of any person to preserve or restore any right or other thing in or connected with a party wall in case of the party wall being pulled down or rebuilt.'

11.1.7 Access

Section 8 of the Act confers a right of access to other property under certain conditions:

> **'8. Rights of entry**
>
> (1) A building owner, his servants, agents and workmen may during usual working hours enter and remain on any land or premises for the purpose of executing any work in pursuance of this Act and may remove any furniture or fittings or take any other action necessary for that purpose.
>
> (2) If the premises are closed, the building owner, his agents and workmen may, if accompanied by a constable or other police officer, break open any fences or doors in order to enter the premises.

(3) No land or premises may be entered by any person under subsection (1) unless the building owner serves on the owner and the occupier of the land or premises—

(a) in case of emergency, such notice of the intention to enter as may be reasonably practicable;

(b) in any other case, such notice of the intention to enter as complies with subsection (4).

(4) Notice complies with this subsection if it is served in a period of not less than fourteen days ending with the day of the proposed entry.

(5) A surveyor appointed or selected under section 10 may during usual working hours enter and remain on any land or premises for the purpose of carrying out the object for which he is appointed or selected.

(6) No land or premises may be entered by a surveyor under subsection (5) unless the building owner who is a party to the dispute concerned serves on the owner and the occupier of the land or premises—

(a) in case of emergency, such notice of the intention to enter as may be reasonably practicable;

(b) in any other case, such notice of the intention to enter as complies with subsection (4).'

11.2 ACCESS TO NEIGHBOURING LAND ACT 1992

11.2.1 Introduction

The *Access to Neighbouring Land Act* 1992 was enacted to overcome the problem of the landowner who wished to repair his property but needed access to his neighbour's land in order to do so, or at least to do so without undue expense. In the absence of any common law or statutory right to enter upon the neighbour's land for the purpose of executing the repairs, it was not possible to force a neighbour to permit access to his land – often the best that could be hoped for was to be charged a ransom sum for the privilege. To a degree, the Act has addressed this sort of difficulty. It applies to England and Wales, save for Crown land, but does not apply to highways.

11.2.2 Scheme of the Act

The Act operates by permitting the county court, upon an application, to make an 'Access Order'.

11.2.3 Making an application

The application (which must be made in accordance with *CPR Practice Direction* 56, para. 11) can be made by any person who, for the purpose of carrying out works to any land, desires to enter upon adjoining or adjacent land, and needs but does not have the consent of some other person to that entry (section 1(1)). It can, therefore, be made by the person who is intending to carry out the work (for example a contractor), rather than the owner of the land upon which the work is to be done.

11.2.4 The test for an Access Order

The Court will only make an Access Order if it is satisfied of two matters.

The first matter is that the works are reasonably necessary for the preservation of the whole or any part of the dominant land (s. 1(2)).

An applicant will certainly satisfy this test if he can demonstrate that it is reasonably necessary to carry out any 'basic preservation' works to the land, which are defined (s. 1(4)) to mean:

> '(a) the maintenance, repair or renewal of any part of a building or other structure comprised in, or situate on, the dominant land;
>
> (b) the clearance, repair or renewal of any drain, sewer, pipe or cable so comprised or situate;
>
> (c) the treatment, cutting back, felling, removal or replacement of any hedge, tree, shrub or other growing thing which is so comprised and which is, or is in danger of becoming, damaged, diseased, dangerous, insecurely rooted or dead;
>
> (d) the filling in, or clearance, of any ditch so comprised.'

Works which do not fall within the definition of 'basic preservation works' may nevertheless be works reasonably necessary for the preservation of any land, and so pass the first test.

The fact that works incidentally involve some alteration, adjustment or improvement to the land, or the demolition of the whole or any part of a building or structure on the land, will not necessarily disqualify them (s. 1(5)).

The Act also provides (s. 1(6)) that where any works are reasonably necessary for the preservation of the whole or any part of the land, the doing to the land of anything which is requisite for, incidental to, or consequential on, the carrying out of those works shall be treated for the purposes of this Act as the carrying out of works which are reasonably necessary for the preservation of that land, and section 1(7) makes similar provision in relation to inspection for the purposes of assessing and planning any works.

The second matter is that the works in question cannot be carried out, or would be substantially more difficult to carry out, without entry onto the other land.

However, even if the Court is satisfied of these two matters, it will nonetheless decline an Access Order if satisfied that if it made the order:

> '(a) the respondent or any other person would suffer interference with, or disturbance of, his use or enjoyment of the servient land; or
>
> (b) the respondent, or any other person (whether of full age or capacity or not) in occupation of the whole or any part of the servient land, would suffer hardship,
>
> to such a degree by reason of the entry (notwithstanding any requirement of this Act or any term or condition that may be imposed under it) that it would be unreasonable to make the order' (s. 1(3)).

11.2.5 What matters are covered by the Access Order?

The Act sets out a number of matters which the Access Order must specify (s. 2(1)), including the works that may be carried out under the Order, the part of the adjoining or adjacent land which may be accessed, and the timing and duration of access.

The Act also provides that the Order may impose terms and conditions upon the access (s. 2(2)–(4)), including a requirement for insurance, and the payment of compensation for loss, injury or damage.

Furthermore, the Order may also require payment of a sum for the privilege of access (s. 2(6)). The sum payable for the privilege of access will be the amount

> '2(5) . . . which appears to the Court to be fair and reasonable, having regard to all the circumstances of the case, including, in particular:
>
> (a) the likely financial advantage of the order to the applicant and any persons connected with him; and
>
> (b) the degree of inconvenience likely to be caused to the respondent or any other person by the entry;
>
> but no payment shall be ordered under this subsection if and to the extent that the works which the applicant desires to carry out by means of the entry are works to residential land.'

'Residential land' is defined in section 2(7) of the Act.

11.2.6 Effect of an Access Order

Section 3 of the Act sets out in detail the effect of an Access Order. In broad terms, the respondent to the application is obliged to permit the applicant and his associates (defined at section 3(7)) to exercise theirs rights and comply with their responsibilities under the Order.

Subject to contrary provision by the Court or the Act, certain rights and responsibilities automatically come with an Access Order (s. 3(2)-(3)).

The Act provides for Access Orders to bind successors in title (s. 4), and for Access Orders and applications for Access Orders to be registered (s. 5).

The Court may order a person who fails to comply with a term or condition of an Access Order to pay damages (s. 6(2)).

11.2.7 Variation of Access Orders

The Court can, after making an Access Order, discharge, vary, suspend or revive any of its terms or conditions (s. 6(1)). An application for the Court to exercise this power can be made by any of the original parties to the application or by any person bound by the Access Order.

11.3 PUBLIC RIGHTS OF WAY

11.3.1 Introduction

The term 'public right of way', used here interchangeably with 'highway', covers a number of types of right of way available for use by the public, including footpaths, bridleways, and byways. In practice, the primary distinction between these species of public right of way is the use to which they may be put. For example, footpaths may be used only by pedestrians, bridleways by pedestrians and riders, and also cyclists, but cyclists must give way to riders and pedestrians.

Public rights of way may arise as a matter of common law, but often (especially in the case of roads carrying vehicular traffic) they owe their existence to legislation, and furthermore, however they may have arisen, they are subject to certain statutory regimes; that is why it is convenient for this section to appear in a chapter dealing with statutory rights.

Whilst a wide range of statutes cover various related areas, the primary provisions are to be found in the *Highways Act* 1980 and are referred to below. The *Highways Act* 1980 makes frequent reference to the 'highway authority', which is defined in section 1(1) of the Act; in practice, outside of Greater London, the highway authority will usually be the local council.

What follows is merely an introduction to this complicated area of the law; for a detailed treatment the reader is referred to *Public Rights of Way and Access to Land,* by Angela Sydenham

11.3.2 Creation of public rights of way

Creation by statute and adoption

The simplest example of creation of a highway is where land is acquired by the highway authority for the construction of a road. Once the road is constructed it is, without more, a public right of way.

Frequently, a private road may be 'adopted' by the local highway authority pursuant to what is known as a 'section 38 agreement' (the reference being to section 38 of the *Highways Act* 1980). Such an agreement may be entered into before or after the construction of the private road to be adopted. Once adopted, the road becomes a highway maintainable at the public expense. Before agreeing to adopt any road, the highway authority will usually insist (as a term of the section 38 agreement) that the road is improved to a certain standard; if the practice were otherwise, those liable for the maintenance of private roads in disrepair could foist an unwanted financial burden upon the taxpayer.

Dedication

A landowner may dedicate a right of way over his land to the public. If accepted by the public (generally, where the dedication is express, acceptance is by the highway authority acting on behalf of the public), the right of way becomes a public right of way.

Dedication and acceptance may be express, or implied. Where a right of way is expressly dedicated, it may be dedicated subject to express reservations of conditions, e.g. subject to the landowner being entitled to place a gate across the right of way.

Implied dedication may arise either at common law or by section 31 of the *Highways Act* 1980. Implied dedication is, in many ways, similar to the process of acquisition of a private right of way through long user, although there are some significant differences – perhaps most significantly that long user alone is insufficient in the absence of an intention on the part of the landowner to dedicate the right of way. Needless to say, whether or not the requisite intention exists is assessed objectively. Indeed, where section 31 of the *Highways Act* 1980 applies, there is deemed dedication if there has been actual use of the way by the public as of right and without interruption for a period of 20 years immediately prior to the date upon which the right of the public to use the way is brought into question. However, this is a presumption which is rebutted where there is 'sufficient evidence that there was no intention during that period to dedicate' (s. 31(1)). This requires some overt act on the part of the landowner, obvious to the public who would be using the way.

R v Secretary of State for Environment, Food & Rural Affairs, ex p Godmanchester Town Council (2007)

In this case the House of Lords considered the meaning of section 31(1) of the *Highways Act* 1980 in the context of what had to be done by a landowner to rebut the presumption of dedication. The House of Lords held that there can only be sufficient evidence that a landowner had no intention to dedicate a path as a public way if he performed overt acts so that the relevant audience, namely the users of the way, would reasonably have understood his intention. Furthermore, that the landowner's intention not to dedicate does not need to be continuously manifested throughout the 20-year period to which the section relates – merely at some point during that period.

11.3.3 Ownership and maintenance

If created formally, or prior to 16 December 1949 (*National Parks and Access to the Countryside Act* 1949, ss. 47 and 49), a public right of way is likely to be maintainable at the public expense.

Where a public right of way is maintainable at the public expense, the surface of the way generally vests in the highway authority (*Highways Act* 1980, s. 263(1)).

Where a public right of way is not maintainable at the public expense, the property in the right of way remains vested in the landowner, although the highway authority will have powers of control over the highway, and this control may be relevant in establishing liability in nuisance where, for example, a highway has become impassable.

There is a rebuttable presumption that where a public right of way separates land belonging to two different owners, each owner owns the land up to the midpoint of the way (of course, in the case of a right of way maintainable at the public expense, the surface of the way will be vested in the highway authority).

11.3.4 Nature of a public right of way and incidental rights

Like a private right of way, a public right of way is a right to pass and repass over a defined route. Unlike private rights of way, there need not be a dominant tenement and so, in theory at least, a public right of way need not lead anywhere.

There are also a number of activities incidental to this right which are permitted. It is clear that these include parking a car, resting, stopping for a meal. However, the precise boundaries of what are permitted as incidental activities are not clear.

Hickman v Maisey (1900)

In this case the Court of Appeal held that walking backwards and forwards on a stretch of highway about 15 yards in length over a period of an hour and a half for the purpose of observing the performance of racehorses in training on adjacent land, and taking notes, exceeded the ordinary and reasonable user of a highway to which the public were entitled. The result was that the defendant, not having authority to be on the highway for this purpose, was guilty of trespass.

11.3.5 Extinguishment and re-routing of public rights of way

Generally

A public right of way cannot be lost through lack of use. Neither is it possible to acquire title to a public right of way through adverse possession insofar as to do so would restrict the public's right to pass and repass.

Where a public right of way has physically ceased to exist (e.g. through coastal erosion), then it will be lost. That said, it may not always be clear as to when a right of way has physically ceased to exist.

R v Inhabitants of Greenhow (1876)

The inhabitants of Greenhow were indicted for non-repair of a highway for which they were responsible. The highway ran along the slope of a hill several hundred feet above a valley. Two landslips occurred on the slope of the hill and part of the highway was carried away into the valley and its place filled with debris. That debris was in parts some 25 feet above the level of the old highway, and elsewhere two feet below it. The line of the old highway was known, and there was evidence that it was practicable to form a new road along the old track. On the facts, the Court held that there was no evidence of destruction of the highway such as to exempt the parish from liability to repair it.

By statute

Statutes may provide for the extinguishment ('stopping-up') or the re-routing of public rights of way. The relevant statute should be consulted in any given case for the detail of what is and is not permitted. For example, in some cases it is possible to effect a partial extinguishment (e.g. of vehicular rights).

Of particular note are the powers appearing in the *Highways Act* 1980 at sections 116, 118 and 119. Section 116 provides for magistrates to order the stopping-up or diversion of a highway (other than a trunk road or a special road). A diversion can only be ordered with the written consent of every person with a legal interest in the land over which the new route will pass. Whilst persons affected by stopping-up and diversion orders can make representations, there is no provision for compensation for any person affected by the order.

Sections 118 and 119 provide, respectively, that, subject to certain limitations, councils may stop-up or divert footpaths and bridleways. Section 121(2) makes provision, by reference to section 28, for payment of compensation in certain circumstances.

11.3.6 Other public rights

Other than the rights referred to above, statutory rights for the benefit of the public, or specific classes of person, include rights relating to commons (see the *Law of Property Act* 1925, the *Commons Registration Act* 1965 and the *Commons Act* 2006), and rights of access to mountains, moor, heath, down and registered common land (see Part I of *Countryside and Rights of Way Act* 2000).

11.4 TELECOMMUNICATIONS

11.4.1 Statutory rights in relation to telecommunications: the legal basis for such rights, and who can acquire them

The Electronic Communications Code ('the Code'), contained in Schedule 2 of *Telecommunications Act* 1984 (as amended by Schedule 3 to the *Communications Act* 2003), provides a scheme of statutory rights with a view to facilitating the construction and maintenance of communications networks.

Section 106 of the *Communications Act* 2003 provides that OFCOM may direct that the Code applies (with or without qualification) to a person (an 'operator') for the purposes of provision by that person of an electronic communications network, or of a system of conduits which are available or to be made available for use of persons providing such networks.

The Code is dealt with in more detail below. Whilst it confers rights directly upon operators, it should be noted that in appropriate circumstances, potential subscribers to an operator's network are entitled to require the operator to exercise its rights under the code, in default of which, to exercise those rights themselves (para. 8 of the Code). Furthermore, those with interests in adjoining land have certain rights in respect of requiring alterations to apparatus (see below at 11.4.4) and compensation (see below at 11.4.5).

11.4.2 What rights can be acquired

The purpose of the legislation is to facilitate the provision of the operator's network. It is therefore unsurprising that the categories of rights which an operator can acquire in respect of land belonging to or occupied by others are very wide indeed (para. 2(1) of the Code):

- the right to execute works on the land for and in connection with the installation maintenance adjustment repair or alteration of electronic communication apparatus;
- the right to keep electronic communication apparatus installed on under or over the land; and

- the right to enter land to inspect electronic communication apparatus on, under, over the land, or even located elsewhere.

There are other rights which can be acquired, set out in other sections of the Code, but in the main they are likely to concern public bodies and utility companies and so will not be dealt with in detail here; those rights include:

- rights in respect of public streets (para. 9 of the Code);
- rights in respect of flying lines (para. 10 of the Code);
- rights in respect of tidal waters or lands (para. 11 of the Code);
- rights in respect of crossing land used in connection with railways, canalways and tramways ('linear obstacles') (paras 12, 13 and 14 of the Code);
- rights in respect of tree-lopping (para. 19 of the Code).

11.4.3 How rights may be acquired and challenged

Wayleave agreements

Broadly speaking, the scheme of the code is that in order to execute works on any land or to keep equipment on land or to enter upon land to inspect apparatus, the operator must have the agreement in writing of the occupier for the time being of the land (para. 2(1)). This is usually referred to (and will be referred to here) as a 'wayleave agreement'.

The occupier is not, of course, obliged to give his agreement, but if he does not the operator may have recourse to other provisions of the code (see below)

Who is bound by a wayleave agreement?

Generally, where a person gives their agreement in writing, that wayleave agreement will bind anyone who derives their interest from them (para. 2(4) of the Code).

Furthermore, if an occupier under a lease for a year or more enters into a wayleave agreement for the provision of a telecommunication service to himself, that agreement will bind not just that occupier, but every other person who has an interest in the land as though they themselves had entered into it – but only for so long as the land continues to be occupied by either the occupier or a person who has agreed in writing to be bound by the wayleave agreement or by a person whose interest is carved out of or derives from such a person. So, for example, a freeholder will be bound by the wayleave agreement granted only by his tenant under a five-year lease, while the tenant remains in occupation. When the lease expires, if the tenant vacates and the freeholder goes into occupation, he will not be bound by the wayleave agreement.

Any person not for the time being bound by a wayleave agreement (typically a freeholder or intermediate lessee) can require the operator to restore the land (para. 4(2) of the Code). This is dealt with in more detail below at paragraph 11.4.4. For this reason it is usual for an operator to seek at the outset the agreement not merely of the occupier but also the freeholder and any other person who might have rights pursuant to para. 4(2) of the Code.

A wayleave agreement does not need to be registered (para. 2(7) of the Code) at the Land Registry and can be enforced by every person for the time being bound by it (para. 2(5)).

Where no wayleave agreement is agreed

In the event that a wayleave agreement cannot be negotiated, there are mechanisms available to the operator by which he can, in effect, compulsorily purchase the right he seeks.

In respect of general rights (para. 2 of the Code) this process commences with the service by the operator of a notice on the person from whom the right is sought or whose interest the operator seeks to bind (para. 5(1) of the Code). In the absence of consent within 28 days the operator may apply to the court for an order granting the operator the right in question or providing for the person or land to be bound by the right and dispensing with the person's consent (para. 5(2) of the Code).

The Court will make an order if (and only if) it is satisfied that any prejudice caused by the order:

- can be compensated for by money; or
- is outweighed by the benefit accruing from the order to the persons whose access to the network or services will be secured by the order.

In considering those matters the Court must have regard to all of the circumstances and to the principle that no person should unreasonably be denied access to an electronic communications network or services (para. 5(3) of the Code).

Any order made may be subject to terms and conditions, and such terms and conditions must include appropriate provisions for ensuring that the least possible loss and damage is caused by the exercise of the right.

It follows from the above that the situations in which it will be possible to defeat entirely an application by an operator are likely to be rare, but there may well be considerable scope for arguing about appropriate terms and conditions, and, of course, compensation, which is dealt with further below at paragraph 11.4.5.

11.4.4 Removal of apparatus

Any person who is entitled to require removal of telecommunications apparatus (e.g. because they are not bound by a wayleave agreement) may only exercise that right in accordance with para. 21 of the Code. That paragraph provides a mechanism for the exercise of the right, but includes considerable safeguards for the operator.

The party seeking removal must serve the operator with a notice. Only if the operator fails to serve a counter-notice within 28 days is the party entitled to enforce removal.

A party entitled to enforce removal may apply to the Court for authority to carry out the removal. The advantage of this is that any expenses incurred by the party in carrying out the removal may be recovered from the operator, and the Court can also order that the party is entitled to sell the apparatus removed and use the proceeds to defray the cost of removal.

If an operator serves a counter-notice, then that counter-notice must specify one or both of the following:

(a) that the party seeking removal is not entitled to require removal; and/or

(b) the steps which the operator proposes to take for the purpose of securing a right as against that person to keep the apparatus on the land.

In the case of (b), this is likely of course to include utilisation of the procedure under para. 5 of the Code described above. In respect of apparatus already installed, para. 6 of the Code allows the court to grant the operator temporary rights whilst waiting for a court determination under paras 5 or 21 of the Code.

Other provisions of the Code deal with the rights of persons affected to object to overhead apparatus (para. 17 of the Code). Operators are obliged to affix to apparatus notices giving their name and an address at which notice of such an objection can be served (para. 18 of the Code).

A person with an interest in land on which apparatus is installed, or adjacent to such land, may give notice to an operator requiring alteration to the installation on the ground that the alteration is necessary to enable them to carry out a proposed improvement of the land (para. 20 of the Code). This is so regardless of any wayleave agreement which may otherwise be binding upon them (para. 20(1) of the Code). As with the rights under para. 21 of the Code relating to removal of apparatus, the operator may serve a counter-notice and ultimately the court will determine the matter.

It should also be noted that it is expressly provided (para. 3 of the Code) that an operator is not entitled to exercise a right acquired under paras 2, 9, 10 or 11 of the Code, so as to interfere with or obstruct any means of entering or leaving any other land unless the occupier for the time being of that other land has agreed or is bound by a right entitling the operator to interfere with or obstruct the means of entering or leaving that other land.

11.4.5 Compensation

Obviously, in any wayleave agreement reached between an occupier of land and an operator, there is likely to be a provision for the compensation of the occupier of the land. Frequently wayleave agreements will be made not just with the occupier(s) for the time being of land but with all of those with an interest in the land. Where however this is not the case, the Code makes provision for compensation to be paid to certain persons with an interest in the land and who may become occupiers by reason of that interest at some point in the future (para. 4 of the Code). Any question as to entitlement to compensation is dealt with by the Lands Tribunal.

In the event that a wayleave agreement is not concluded by consent and the operator exercises his rights pursuant to paragraph 5 of the Code, the Court may fix financial terms (para. 5(4) of the Code), and such terms must include (para. 7(1) of the Code):

(a) such terms with respect to the payment of consideration in respect of the giving of the agreement, or the exercise of the rights to which the order relates, as it appears to the Court would have been fair and reasonable if the agreement had been given willingly and subject to the other provisions of the order; and

(b) such terms as appear to the Court appropriate for ensuring that that person and persons from time to time bound by the rights conferred are adequately compensated (whether by the payment such consideration of otherwise) for any loss or damage sustained by them in consequence of the exercise of those rights.

The Court must also take into account the prejudicial effect (if any) of its order on the affected party's enjoyment of or interest in land other than the land which is the subject of the rights being conferred (para. 7(2) of the Code).

Mercury Communications Ltd v London & India Dock Investments (1993)

The applicant operator asked the Court to dispense with the respondent's consent in respect of proposed new ducts under a private road owned by the respondent. It was common ground that the respondent would not suffer any loss or damages by reason of the order (para. 7(1)(b) of the Code) and that there would be no prejudicial effect on other land that they owned (para. 7(2) of the Code). The only question therefore was the amount to be awarded under para. 7(1)(a) of the Code. It was also common ground that the valuation date should be the date of the court order.

HHJ Hague QC, sitting at Mayor's & City County Court, held that the following principles applied to assessment under para. 7(1)(a) of the Code:

- The Court was required to determine what would be fair and reasonable had agreement been given willingly, and although the market value was an obvious starting point, it may not necessarily produce a figure that was fair and reasonable.
- Compulsory purchase principles are not relevant.
- It is not appropriate to have regard to the benefits which would be conferred upon the operator by the provision of the facility save perhaps in a case in which a single capital payment is to be made and the amount of the benefit can be easily quantified.

The Learned Judge assessed the compensation at an annual sum of £9,000 by reference to comparables and having regard to the bargaining position of the parties.

There is also provision for compensation to be paid for the adverse effect of the exercise of rights under the Code on neighbouring land (para. 16 of the Code). The compensation payable by an operator is expressed to be payable under section 10 of the *Compulsory Purchase Act* 1965 as if that section had effect in relation to injury caused by the exercise of the rights as it has effect in relation to injury caused by the execution of works on land that has been compulsorily purchased.

11.5 WATER

11.5.1 Statutory rights in relation to the transportation of water: the legal basis for such rights, and who can acquire them

The relevant statutes are the *Water Industry Act* 1991 and the *Water Resources Act* 1991 (as amended).

Under the *Water Industry Act* 1991, the Secretary of State may appoint or authorise the appointment of a company as the water undertaker or sewerage undertaker for any area in England and Wales. Such undertakers are granted wide ranging powers under the Act, which are dealt with below at 11.5.2.

The *Water Resources Act* 1991 confers upon the National Rivers Authority (whose functions are now carried out by the

Environment Agency) powers which are almost identical to those granted to statutory undertakers by the *Water Industry Act* 1991.

11.5.2 What rights are available

Under the *Water Industry Act* 1991, the rights available to statutory undertakers include:

- Subject to authorisation by the Secretary of State, rights of compulsory purchase of land, including the creation of new rights as well as the acquisition of new ones, and also the extinguishment of existing rights (s. 155(2)).
- The power to lay pipes in a street and to carry out ancillary activities including inspection, maintenance, alteration and repair (s. 158).
- The power to lay pipes in other land and to carry out ancillary activities including inspection, maintenance, alteration and repair (s. 159).
- The power to carry out works for sewerage purposes (s. 160) and to deal with foul water and pollution (s. 161).
- Entry onto land for the purposes of exercising its powers or ascertaining whether and/or how to exercise any such powers (s. 168), and for a variety of other purposes (ss. 169–172).

Similar powers are set out in the *Water Resources Act* 1991, sections 154–174.

11.5.3 Challenging the rights and compensation

In the case of the exercise of rights of compulsory purchase under section 155(1)–(2) of the *Water Industry Act* 1991, and section 154(1)–(2) of the *Water Resources Act* 1991, the *Acquisition of Land Act* 1981, and the *Compulsory Purchase Act* 1965, are applied, subject to some modifications (section 155(2)3–6 as regards the *Water Industry Act* 1991, and section 154(2)3–6 as regards the *Water Resources Act* 1991).

In the case of statutory undertakers, Schedule 12 to the *Water Industry Act* 1991 imposes obligations on them to minimise damage caused in the exercise of certain powers and also obligations as to compensation. Schedule 21 to the *Water*

Resources Act 1991 makes similar provision in respect of the exercise of powers under that Act.

Section 181 of the *Water Industry Act* 1991 also provides for the investigation by the Director General of Water Services of complaints by the public about the carrying out of works on private land by statutory undertakers. He can direct statutory undertakers to pay up to £5,000 compensation. Section 182 sets out a code of practice with respect to work on private land.

11.6 GAS

11.6.1 Statutory rights in relation to the transport of gas: the legal basis for such rights, and who can acquire them

The relevant statute is the *Gas Act* 1986 (amended by the *Gas Act* 1995 and the *Utilities Act* 2000). Section 7 of the Act provides that the Gas and Electricity Markets Authority may grant licences authorising a person, within certain parameters, to convey gas through pipes to premises or pipe-line systems.

When acting for purposes connected with the carrying on of activities authorised by their licence, or the conveyance of gas through pipes situated in certain areas, or to or from a country or territory outside of Great Britain, such licence holders are known as 'gas transporters'.

11.6.2 What rights are available?

Schedule 3 provides that, after consultation with the Gas and Electricity Markets Authority, the Secretary of State may authorise a gas transporter to purchase compulsorily any land, including any right over land, and this includes the acquisition of rights over land by creating new rights as well as acquiring existing ones.

Schedule 4 provides gas transporters with the right to execute various kinds of works relating to laying, repairing, altering and removing pipes and other apparatus in or under any street (para. 2(1)), and also authorises incidental works – for example, the opening or breaking up of a street (para. 2(2)). However, these provisions do not authorise the laying down or placing of any pipe or other works into, through or against any building, or in any land not dedicated to the public use (para. 3); the only exception to this is that the rights conferred may be exercised in relation to a street which has been laid out but not dedicated to the public use for the purpose of conveying gas to any premises abutting the street.

11.6.3 Challenging the rights, and compensation

In the case of the exercise of rights of compulsory purchase under Schedule 3, the *Acquisition of Land Act* 1981 and the *Compulsory Purchase Act* 1965 are applied, subject to some modifications – in particular there is an amendment to section 7 of the 1965 Act dealing with the matters to which regard must be had in assessing compensation.

In the case of the exercise of rights under Schedule 4 (see above at 11.6.2), the gas transporter is under a statutory obligation to do as little damage as possible in the exercise of the powers, and to make compensation for any damage done (para. 1(3)).

Regulations provide that in certain cases, compensation will be payable to those who suffer a loss as a result of the exercise of powers under Schedule 4 (the *Gas (Street Works)(Compensation of Small Businesses) Regulations, SI 1996/491*).

The powers referred to at 11.6.2 above include the power to erect in any street one or more structures for housing any apparatus, but only with the consent, which shall not be unreasonably withheld, of the street authority (para. 2). In the event of a dispute as to whether or not consent is being unreasonably withheld, this is to be determined by a single arbitrator appointed by the parties, or in default of agreement by the Gas and Electricity Markets Authority.

11.7 ELECTRICITY

11.7.1 Statutory rights in relation to the supply of electricity: the legal basis for such rights, and who can acquire them

Like other statutes dealing with utilities, the scheme of the *Electricity Act* 1989 (amended by the *Utilities Act* 2000) is to permit the Secretary of State to grant licences to providers authorising the generation transmission and supply of electricity to any person in the provider's authorised area (s. 6).

Those licences may provide that certain rights, which are set out in Schedules 3 (compulsory acquisition of land) and 4 (other powers and provisions) to the 1989 Act, are available (or available subject to qualifications) to the holder of the licence. In other words, the potentially available rights themselves are to be found in Schedules 3 and 4, but the terms of any given licence will set out the extent to which that particular licence holder has the benefit of those rights.

11.7.2 What rights are available?

The rights potentially available to licensees include:

- Subject to authorisation by the Secretary of State, rights of compulsory purchase of land, including by the creation of new rights as well as the acquisition of new ones (Sched. 3, para. 1).
- The right to carry out works to install lines or plant, or structures for housing or covering lines or plant, and to carry out ancillary activities including inspection, maintenance, alteration and repair (Sched. 4, paras 1 and 2).
- A wayleave allowing the licence holder to instal and keep an electric line on, under or over any land (Sched. 4, para. 6).
- Felling or lopping of trees in proximity to electric lines or installations (Sched. 4, para. 9).
- Entry onto land for the purposes of ascertaining whether the land would be suitable for use for any purposes connected with the carrying on of authorised activities (Sched. 4, para. 10).

11.7.3 Challenging the rights and compensation

In the case of the exercise of rights of compulsory purchase under Schedule 3, the *Acquisition of Land Act* 1981 and the *Compulsory Purchase Act* 1965 are applied, subject to some modifications (Sched. 3, paras 5–14). When land or an interest in land is compulsorily purchased the affected owner will generally be entitled to compensation.

As regards wayleaves, reference should be had to Sched. 4, para. 6. In broad terms, where it is necessary or expedient for a licence holder to install and keep installed an electric line on, under or over any land, the licensee may give the occupier of the relevant land notice requiring him to grant the wayleave within a specific period of time (which cannot be less than 21 days). If the occupier is unwilling to make the grant, then the licensee can make an application to the Secretary of State.

The Secretary of State cannot grant the application if the line is to be installed on or over land covered by a dwelling, or land which will be so covered on the assumption that any planning permission which is in force is acted on (para. 6(4)). The definition 'dwelling' is quite wide (see para. 6(8)) and includes a garden.

Before granting a wayleave, the Secretary of State will afford the occupier and, if different, the owner, the opportunity of being heard by a person appointed by the Secretary of State. The process is governed by the *Electricity (Compulsory Wayleaves)(Hearings Procedure) Rules* 1967 (as amended by the *Electricity Act 1989 (Consequential Modifications of Subordinate Legislation) Order* 1990.

If a wayleave is granted by the Secretary of State it is usually subject to certain terms and conditions. This invariably includes payment of compensation to the occupier and/or owner of the land in respect of the grant.

A wayleave does not need to be protected by registration.

Index

The *Case in Point* series

The *Case in Point* series is a popular set of concise practical guides to legal issues in land, property and construction. Written for the property professional, they get straight to the key issues in a refreshingly jargon-free style.

Areas covered:

Negligence in Valuation and Surveys
Item code: 6388
Published: December 2002

Party Walls
Item code: 7269
Published: May 2004

Service Charges
Item code: 7272
Published: June 2004

Estate Agency
Item code: 7472
Published: July 2004

Rent Review
Item code: 8531
Published: May 2005

Expert Witness
Item code: 8842
Published: August 2005

Lease Renewal
Item code: 8711
Published: August 2005

VAT in Property and Construction
Item code: 8840
Published September 2005

Construction Adjudication
Item code: 9040
Published October 2005

Dilapidations
Item code: 9113
Published January 2006

Planning Control
Item code: 9391
Published April 2006

Building Defects
Item code: 9949
Published July 2006

Contract Administration
Item Code: 16419
Published March 2007

Construction Claims
Item Code: 16978
Published December 2007

If you would like to be kept informed when new *Case in Point* titles are published, please e-mail **rbmarketing@rics.org**

All RICS Books titles can be ordered direct by:

☎ Telephoning 0870 333 1600

Online at http://www.ricsbooks.com

E-mail mailorder@rics.org